Lecture Notes in Artificial Intelligence 1699

Subseries of Lecture Notes in Computer Science
Edited by J. G. Carbonell and J. Siekmann

Lecture Notes in Computer Science

Edited by G. Goos, J. Hartmanis and J. van Leeuwen

Springer

Berlin
Heidelberg
New York
Barcelona
Hong Kong
London
Milan
Paris
Singapore
Tokyo

Sahin Albayrak (Ed.)

Intelligent Agents for Telecommunication Applications

Third International Workshop, IATA'99
Stockholm, Sweden, August 9-10, 1999
Proceedings

 Springer

Series Editors

Jaime G. Carbonell, Carnegie Mellon University, Pittsburgh, PA, USA
Jörg Siekmann, University of Saarland, Saarbrücken, Germany

Volume Editor

Sahin Albayrak
Technische Universität Berlin, DAI-Laboratory
Franklinstraße 28/29, D-10587 Berlin, Germany
E-mail: sahin@cs.tu-berlin.de

Cataloging-in-Publication data applied for

Die Deutsche Bibliothek - CIP-Einheitsaufnahme

Intelligent agents for telecommunication applications : third
international workshop ; proceedings / IATA '99, Stockholm, Sweden,
August 9 - 10, 1999. Sahin Albayrak (ed.). - Berlin ; Heidelberg ;
New York ; Barcelona ; Hong Kong ; London ; Milan ; Paris ;
Singapore ; Tokyo : Springer, 1999
  (Lecture notes in computer science ; Vol. 1699 : Lecture notes in
  artificial intelligence)
  ISBN 3-540-66539-0

CR Subject Classification (1998): I.2.11, C.2, H.4.3, H.5

ISBN 3-540-66539-0 Springer-Verlag Berlin Heidelberg New York

© Springer-Verlag Berlin Heidelberg 1999
Printed in Germany

Typesetting: Camera-ready by author
SPIN 10704559   06/3142 – 5 4 3 2 1 0   Printed on acid-free paper

Preface

The first international workshop on Intelligent Agents for Telecommunications Applications (IATA'96) was held in July 1996 in Budapest during the XII European Conference on Artificial Intelligence ECAI'96. The workshop program consisted of technical presentations addressing agent based solutions in areas such as network architecture, network management, and telematic services. Presentations gave rise to a lively debate on the advantages and difficulties of incorporating agent technology in telecommunications. The proceedings were published by IOS Press providing introductory papers on agent technology as well as telecom applications and services and also papers about appropriate languages and development tools.

The second International Workshop, IATA'98, was held in Paris, in the framework of Agents' World which brought together the principal scientific and technical events on agent technology such as the International Conference on Multi-Agent Systems (ICMAS'98), RoboCup'98 devoted to an international competition between soccer-playing robot teams, and six international workshops. Each workshop focused on specific aspects of agent technology such as databases and information discovery on the Internet (CIA'98), Collective Robotics (CRW'98), Simulation (MABS'98), Agent Theories, Architectures and Languages (ATAL'98), Communityware (ACW'98), and Telecommunications Applications (IATA'98). The proceedings of IATA'98 were published by Springer-Verlag.

Agent technology is a very promising approach to addressing the challenges of modern day telecommunications. The existing world of telecommunications – which is deeply influenced by monopolistic public network operators (PNOs) – is currently changing at a rapid pace. This change is taking place in the technological as well as the regulatory arena. Additionally, market forces on an unprecedented scale are at work. Given this change, it will no longer be sufficient for PNOs to solely provide network infrastructure. The challenge for PNOs consists in evolving to full-service providers. This implies that, on the one hand, an increasingly complex telecommunications infrastructure needs to be managed more efficiently and, on the other hand, that new types of telecommunications services need to be developed and provided. It is particularly such future services that need to satisfy a diverse range of requirements, e.g. personalization, support for user mobility, on-demand combination of different services, offline/online service usage etc..

Agent technology addresses these requirements particularly well as opposed to other technologies, e.g. client-server. A stationary agent can reside on agent platforms "in the net", providing various types of services. Besides being potentially decentralized and cooperative, these stationary service provider agents possess capabilities for issues of security, accounting, and billing etc. On the client side, agent-based services will be requested by means of small, mobile agents which may enable both offline and online service usage. Agent technology is very well supported by the language Java and corresponding Java APIs.

The aim of IATA'99 is to provide a state-of-the-art forum for presenting innovative agent based applications in telecommunications, and for discussing new approaches, new models, and technology trends in both telecommunication and agent related fields.

This volume contains a revised version of the papers selected by the program committee for presentation and discussion at IATA'99.

The book comprises a collection of fourteen papers organized into four groups. Contributions in the first group present *architecture, tools, platform, and languages* for development of agent-based systems for TelCos. This first part starts with a presentation of a toolkit for the realization of agent-based applications in the telecommunications domain.

The second group deals with new approaches for *network management solutions* that can be realized by using agent technology.

The third group shows how *e-commerce* platforms and services can be realized based on agent technology.

The last group's contribution comprises 5 papers. The first one shows how agent technology can realize telecommunication services and how the service provisioning is established based on agent cooperation. The second paper describes the realization of VPN services based on agent technology. The third paper shows how agent technology can be deployed in universal messaging. Finally, the last paper describes the realization of agent-based brokerage.

Acknowledgements

We would like to express our sincere gratitude to all the people who helped to bring about the production of this book.

Nothing would have been possible without the initiative and dedication of the DAI-Lab team at the Technical University of Berlin.

We owe particular gratitude to the members of the program committee for their professionalism and dedication in selecting the best papers for the workshop. We especially thank all contributing authors for choosing IATA'99 to present their research results, and for their diligence and their cooperation in the preparation of this volume.

Hans Schlenker of the DAI-Lab has organized the review process, keeping in touch with the authors and monitoring the submitted contributions and the accepted papers. He did a great job.

Finally, we would like to express our appreciation of the various workshop sponsors:
- Deutsche Telekom
- France Telecom
- Sun Microsystems
- Alcatel
- Siemens AG

August 1999   Sahin Albayrak

Organizing Committee

Chairman

Sahin Albayrak
    Technische Universität Berlin, Germany

Co-Chairmen

Charles Petrie
Center for Design Research, Stanford University, USA

Lars B. Jansson
Sun Microsystems AB, Sweden

Program Committee

Thierry Bouron, France
Yves Demazeau, France
Innes Ferguson, Canada
Tim Finin, USA
Francisco Garijo, Spain
Toru Ishida, Japan
Nick Jennings, UK
Paul Kearney, UK
Danny Lange, USA
Victor Lesser, USA
Divine Ndumu, UK
Hyacinth Nwana, UK
Peter Selfridge, USA
Munindar P. Singh, USA
Katia Sycara, USA
Robert Weihmayer, USA
Frank von Martial, Germany

Table of Contents

Telecommunication Services

JIAC – A Toolkit for Telecommunication Applications

Sahin Albayrak, Dirk Wieczorek

Technical University of Berlin
DAI-Laboratory
Email: sahin@cs.tu-berlin.de, wieczore@cs.tu-berlin.de

Abstract. This paper describes the JIAC (Java Intelligent Agent Componentware) architecture, an open and scalable agent architecture. The telecommunication market of today being in continuos expansion is in need of a platform to access the resulting growing demands. JIAC provides an agent based platform to meet those demands of the telecommunication applications. Agent oriented techniques are considered the adequate solution for the maintenance of networks and the provisioning of services.
JIAC is using the component based approach to build customised agents. To realise telecommunication applications, JIAC relies on electronic marketplaces as the basic structure.
Keywords: Java based agents, electronic market places, e-commerce, telematic services, telecommunication applications, telecommunication service provisioning, mobility support, intelligent agents for telecommunication application and future telematic services, agent architecture

1 Introduction

The telecommunication market is changing very rapidly. On the one hand, the basic technologies for telecommunication in general are developing very fast. On the other hand, the liberalisation of the telecommunication market has urged the existing network operators to open their networks to new suppliers. A result is the network for everyone. The offered network services will be crucial with respect to the competition in the future market. Therefore, the telecommunication market will have to use existing networks efficiently and to control them. Also, the conception of innovative services and the provisioning of these new services will be an important task. These future services will have to meet a number of requirements as described in the introduction of this book.

Agent oriented techniques are a very promising approach to realise future telecommunication applications and telematic services. For their efficient and effective development particular agent platforms are needed. They represent an architecture and other constructs as basic elements as well as suitable tools for the system realisation. Additionally, they typically include agent-oriented tools for the validation of the implemented systems (high-level debugging).

JIAC is such a platform, thus providing an open and scalable agent architecture mainly based on the metaphorical concept of electronic marketplaces. Being very general in its principal structure, arbitrary agent systems in the described domain can be constructed using JIAC.

In the following sections, we first categorise and describe more in detail the type of telecommunication applications and telematic services we have in mind. Subsequently, we derive the obvious demands the applied agent platform has to fulfil in order to support and ease the development of these services. Following that, we go into the details of the JIAC platform. A description of available tools and additional remarks conclude this paper.

2 Requirements on Agent Architectures for Telecommunication Services and Applications

Future telecommunications applications and telematics services will have to satisfy a broad range of user requirements. As the future telecommunications markets will be directly influenced by the available services and applications the service architecture, here particularly the agent architecture, must exhibit a number of technical characteristics. Such characteristics include:

- *scalability*: the same agent architecture should be deployed for both simple and complex functionalities as well as different types of agents, e.g. mobile or stationary. Also, service functionality and components should be easily added and removed during service run-time. Finally, service performance should not degrade in the face of large numbers of users;
- *manageability*: the agent architecture must support management functions (as capabilities) such as configuration, fault and performance management;
- *security*: the agent architecture must offer a basic set of security functions, but also it must support the procurement of additional security functions and services during run-time;
- *accounting*: the agent architecture must support the collection of transaction data at various levels of detail and according to selected accounting models for particular billing purposes;
- *openness*: standard and de-facto standard protocols both for basic communication but also for service co-operation must be supported, e.g. TCP/IP, KQML, FIPA ACL;
- the agent architecture should offer means to *wrap existing legacy systems* in order to protect past investments and to easily add value to existing services;
- services based on the agent architecture should reside "in the network" in order to allow service portability and robustness.

If agents are used for telecommunication applications, especially for service provisioning, it is necessary to have a certain structure for the agents to live in. The structure is provided by the notion of electronic marketplaces: a broad range of agent-based telecommunication applications and services may be realised by using agent platforms modelled after electronic marketplaces.

3 JIAC: The Toolkit for Agent Oriented Telecommunication Applications

JIAC is the abbreviation of Java Intelligent Agent Componentware. It is a framework that supports agent oriented software development.

With the ongoing success of the Java programming language [Java] and its suitability for mobile and platform-independent applications, we observe a growing popularity of Java in the mobile agents area. JIAC has some unique features for the realisation of telematics applications, it is highly scalable, and it especially supports the wrapping of legacy systems with exchangeable I/O components.

JIAC contains an implementation of the concept of an electronic marketplace. With the help of this platform telecommunication applications and future telematic services can be realised. Additionally, the adequate modelling and realisation of e-commerce solutions is supported by certain capabilities of the agent architecture, like accounting and billing, and the marketplace construct itself.

JIAC is a pure Java API implementing component based mobile agents. Apart from distributed, mobile agents, JIAC features a complete set of structures, which are realised by the notion of marketplaces, like accounting, billing, and security for electronic commerce applications or distributed provisioning of telecommunication services.

4 Agent oriented software-engineering

Agents are, like objects, a try to simplify the building of software by breaking down the complexity to a manageable level. Often agents are seen as an extension to the object oriented approach, they are just more complex objects with more flexible interaction. Thus most earlier mentioned AOSE (Agent Oriented Software-Engineering) approaches have their OOSE (Object Oriented Software-Engineering) counterpart, but are also depending on the underlying agent definition or the used agent architecture.

Based on this approaches an AOSE scheme, orienting on the definition, where agents are defined as autonomous specialist in an agent society, will be presented:

- Analysis: Agents and agent types are identified according to their tasks and rolls in an agent society. The problem area is modelled in following three levels:
  1. The agent model describes the agents and their properties. This includes needed and provided information types and services, user interaction and database access.

2. The organisational model describes the structures which make agent interaction possible, and relations between agents which have special dependencies.

3. The interaction model explains the agents communication behaviour. In contrast to the organisational model real interaction needed for task completion and problem solving is modelled.

- Design: Because of the conceptual differences between different agent systems, unlike using OOSE, it is necessary to choose an useful architecture during design phase. Depending on this further enhancement of the analysis models is done.

The design of single agents and their functionality is strongly dependent on the application area and the underlying architecture. In general it involves data structures and methods, which describe the information and services of the agent. AI based architectures include knowledge representation and planing formalisms.

The used agent architecture should provide parts of the organisational structures. This includes agent platforms, which offer communication facilities an migration possibilities for mobile agents and special agents for service management and brokering. Organisation of agents in hierarchies or groups can be forced explicitly by control structures or remain implicitly in the interaction structures.

Protocols, speechacts and ontologies are defined for interaction. The involved agents must be able to understand them.

- Implementation: Implementation also strongly depends on the chosen architecture. In any case the during the design phase developed models must transformed in agent societies and concrete agents with according functionality.

Because of the modularity of agent systems and most agent architectures, like in OOP, often the possibility for reuse of components or whole agents, appears. This includes use of standard agents and components in libraries.

- Evaluation: The high abstraction level of agent programming and the complexity of multi agent systems produce a need for tools to analyse system behaviour. Debugging should be possible on both code and agent level. This includes visualisation of the organisational structure and development of the agent society during runtime and the visualisation of agent interaction on speech act level.

4.1 Derivation of an agent based application development method

Extending the above specified phase of agent oriented software development, for each phase an detailed action plan will be founded. Also each phase is extensive supported by software tools. It's not decided yet if this tools remain separate solutions or are unified into a single integrating tool.

- Analysis: During analysis JIAC base agents are identified according to their tasks and rolls in an agent society. But it remains to the programmer to use a single or more agent base classes, which extend JIAC agents. Basically it is possible to extend every JIAC agent by the addition of capabilities, realised as scripts, depending on their tasks.

1. During the building of the Agent Model basically the identified agents are equipped with capabilities (scripts). A supporting tool AgentBuilder exists.
2. The translation of the organisational model is supported by graphical tools, which allow the distribution of agents on the defined market places.
3. The Interaction model is build by definition of the communication protocol. This includes the speechacts and their temporal order and complete transaction models.

- In the design phase the above models are verified in terms of quality in combination with JIAC. Interaction protocols are defined using a ProtocolBuilder tool. Speech act types and their contents are defined in this phase as well.

- Implementation: The implementation is responsible for conversion of above models in JIAC agents. Mostly one or more agent base classes, which the concrete identified agents extend, are used. Implementation of scripts concludes this phase.

- Evaluation: Because of the complexity of multi agent systems there is a need for tools, which allow system behaviour analysis. This is done by AgentMonitor, which is a complete agent oriented debugger

5 Agents and Marketplaces

The JIAC environment offers two main actors: **agents** and **marketplaces**.

In JIAC, agents are software entities with a reactive and asynchronous task processing. Agents pursue their goals autonomously, including the migration to remote marketplaces there to call on offered services they need to fulfil their tasks. However, the user of an agent can be notified by the agent from the remote location on every autonomous migration. So the user can be in continuous control of the agent.

In JIAC, all agents communicate with each other by the use of speech acts. The implementing protocol follows the proposed standard KQML (Knowledge Query and Manipulation Language [KQML]).

Apart from these, speech acts in JIAC may contain specific additional attributes extending the KQML standard. For instance, these attributes specify optional encryption algorithms used, id's of accounts possibly associated costs of the speech act have to be charged to etc.

JIAC supports the following speech act standard KQML types (called *performatives*): **Tell, Untell, AskOne, Sorry, Reply, Evaluate, Register**, and **Unregister**.

To describe the capabilities of agents we adopt the metaphor of a service. Since our agents are implemented using the reactive agent architecture approach, an agent 'reacts' to events from its environment by invoking immediate actions using a fixed pattern-response table. Such a pattern is called a *service* in JIAC. To address services comprehensively, they can be collected into groups called *categories*.

In order to meet the demands of applications, JIAC provides two basic types of agents:

- **stationary agents,** which exist in marketplaces and offer services. They are constantly assigned to a certain marketplace. The following types of agents may be considered stationary agents:
  - marketplace manager agents (see below)
  - service provider agents
  - content provider agents
- **mobile agents,** which migrate from marketplace to marketplace. Such agents are defined as software programs, typically written in an interpreted language, that have the special functionality to move from one execution context (computer) to another.

These mobile agents might visit several marketplaces in order to find a suitable service provider agent. **Mobility** of agents is one of the features which are supported through the use of marketplaces. This allows agents to move around and enables them to perform their tasks locally, in the locations where the involved resources/entities are located, rather than "shouting" requests across the network in order to access resources remotely. Consequently, the underlying concept is often referred to as "Remote Programming" (RP). The combination of electronic marketplaces and mobile agents might help to reduce online cost: Instead of a continuous network connection, a service call consists of an initial connection to transport the mobile agent including any input constraints to the server marketplace, an intermediate period of computation time without online access, and a second connection used to transmit back the agent together with the results. This service structure allows the customer to have a low-bandwidth connection as e.g. provided by a cellular telephone. Because in a retrieval and filtering process on the server itself the information is cut down by the agent, the actual amount of transmitted data is minimised. Having the majority of computing done on the server, the customer needs not to have large processing capabilities.

The difference between mobile code and a plain message is that a user can send 'intelligence' to the destination to act according to the wishes of the user. That is, by an agent we send information *and* the means to handle it rather than sending the information alone.

Following, we focus on the context our agents live in: *electronic marketplaces*.

In simple terms, *JIAC-Marketplaces* are software environments, where several agents can gather to offer their specific services. The marketplaces are organised in tree-like hierarchies, and each marketplace is administrated by a special *manager-agent*. Manager agents offer an interface for the use of the services provided by the agents within their corresponding marketplace. Another aspect that is covered by manager agents is to provide the facilities for migration of mobile agents. Furthermore, they take care of infrastructure issues on a marketplace like *logging* and *security*.

Marketplaces are not limited to one host. It is straightforward to distribute a marketplace over several hosts and, due to Java, even across several hardware/operating system platforms.

Generally, agents (especially service provider agents) have to be localised in the network in order to be contacted or used. As agents reside on marketplaces, this problem can be reduced to the localisation of marketplaces. This addressing can be organised in a flexible way if marketplaces are not only stand-alone structures but inter-linked in a tree-like hierarchy. This allows us to implement a uniform directory service based upon traversing of these trees.

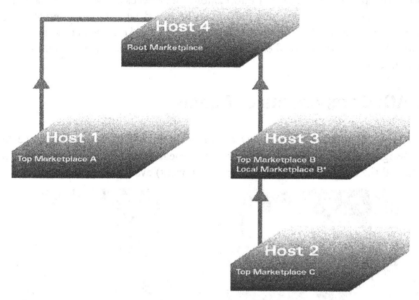

Figure 1: A Hierarchy of Marketplaces

The residents on these marketplaces are stationary and mobile agents. A special stationary agent is the manager agent taking care of the marketplace.

6 Brokering – In Search of Services

The search for a service provider agent begins within the service requesting agent itself – the request has its origin in one component of the agent, so another component within that same agent can possibly provide the requested service – and continues in the direction of the root manager. Initially, the service requesting agent looks through its own components to find a provider for the asked service. If that does not succeed, the request is given to the superior agent. This agent searches his own components and, in the case of a manager agent, asks all registered agents. If no suitable agent can be found, the request is given once more to the superior agent. This process moves on, until no superior can be found anymore, or a service provider is finally found. In the first case a failure message is generated and in the second case the service is requested.

It may happen, that the whole hierarchy of manager agents is being searched through. Since that is a very costly process, the data is cached to allow the efficient retrieval of service providers.

The process of searching for a service provider can be pruned earlier, if the knowledge of the manager agents about the categories of services is taken into account. These categories are known all the way up through the hierarchy to the root manager. So that this may act as a rather unspecified hierarchical inventory of services. The complete description of agent services is stored locally in the particular agent (service provider). This set of information can be accessed with a dedicated tool to present it nicely to the user. So agents can be asked what kind of services they provide, or, more generally, the service categories of a marketplace can be requested. Additionally, the information whether the service is commercial or free, is transferred to the user, too.

7 JIAC: Components of Agents

A JIAC agent comprises several *components*. These components are exchangeable by the designer as well as by the agent itself. This means, an agent is able to change its 'equipment' (i.e.communication or security component) even during runtime.

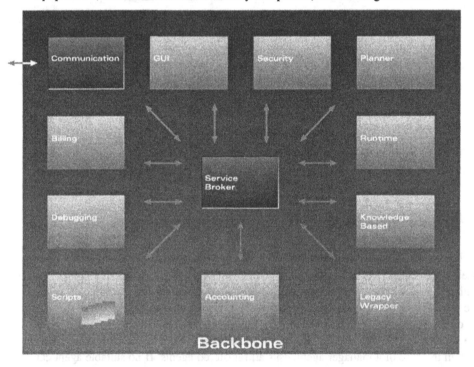

Figure 2: Internal View of an Agent

Featuring the aspect of a construction set approach, JIAC allows to agents to scale themselves by dynamically adding or removing components or scripts. That enhances the functionality of both agents and marketplaces since new services can be added or updated anytime without the need to stop the marketplace or the agent. As shown in Figure 2, there are however some significant components of the system like a service broker or a communication component which have to be present in the agent in order to run properly.

Each JIAC component is out of an open set of components and encapsulates a certain semantic aspect of the agent. The components are plugged into a message bus that arranges the scheduling of the internal message traffic of the agent. So the message bus forms the most basic part of an agent, its backbone (cf. Fig. 3).

This component based agent architecture brings all of the known advantages of component based software design [Gamma, Helm, Johnson, Vlissides, 1994], as better reusability of software, more reliable and secure systems and faster realisation of the software.

Each component has three essential aspects:
- Support of both **persistence** and **migration** of the agent.
- **Integration** into the agent. This means the interface that the component has to implement.
- **Concurrency**, which is realised by the fact that each component is a concurrent thread inside the agent.

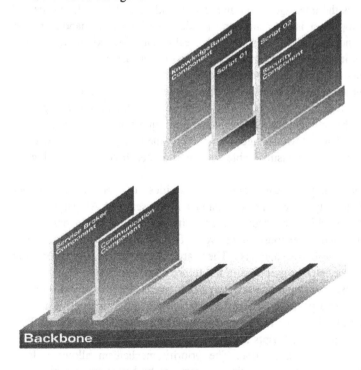

Figure 3: Components of an Agent

As shown in Figure 3, the component-based architecture allows an easy scaling. Each component can have its capabilities registered as services. Each agent is required to have at least a communication component and a service broker component, with which the agent can offer a basic functionality. By adding more components, the functionality or capabilities of an agent are extended. This concept can be compared to the motherboard of a computer. This motherboard has a basis which is used to extend the functionality by adding components. Concerning the JIAC architecture, there are a backbone component and two basic components which are responsible for communication and service brokering, completing the basic functionality of an JIAC agent just like a computer does in combination of its parts.

Reflecting the several aspects of an agent, there is a set of possible components for agents in JIAC:

- The **ServiceBroker** looks up for a certain **service,** given a certain speech act. The service broker can be viewed as a kind of name server for service offering agents. This component implements a special brokering approach. The services of components are grouped by similar services. For each agent component, only the best fitting service is used. Multicasting is done between components, whose services have the same goodness. If a consumer implemented component is inside the set of accepting components, then only the internal order is decisive.

- The **Communication Component** represents the communication facility to other agents and includes features like the communication protocol (cf. Sec. 9), etc. This component realises the inter agent communication with speech acts. It consists of two layers. The top layer is a machine independent mechanism to handle speech acts, and the second layer is the interface to the real machine the communication is running on. This interface is implemented by a so called driver. A driver is the standard interface to a specific communication platform.

- An **I/O Component** encapsulates the complete access to other non-agent systems such as databases, legacy systems (e.g. HTTP server access), etc. This component realises a special driver to access the legacy system in question. This driver can be updated on demand to allow the agent to adapt to new requirements. Also an intelligent converting mechanism is able to translate speech acts into the legacy interface definitions.

- A **Planner Component** encapsulates the abilities of the agent to act autonomously. Currently there is no real planner in the sense of AI. That means, this is a component which can decide what script out of a possible number of scripts is best executed in response to an event. The necessary information about scripts is located in the knowledge base. These are priority values, execution time and cost factors. The later ones should in the future adapt over time. In general, a speech act can cause more than one script to be launched, if this is allowed by the programmer. It is also possible to register a script for different speech acts of services. As mentioned, there is only a very basic planner at the moment. This planner sorts the appropriate scripts according to their priority and consecutively runs the script with the highest one. The priority mechanism allows a later shadowing of low-quality implementations of services by better ones without any need to remove the low-quality implementation.

- An **Accounting Component** logs all actions of the agent and all services it calls on. The accounting component collects data for each commercial script. Commercial scripts have a fixed valued, the execution costs, associated with it. The accounted data is used later to derive a bill. Whenever such a commercial script is initiated, an account is opened automatically and the cost value is added to the account. If speech acts (or service requests) with a valid account number are sent by this script a pending bill counter is increased. Incoming bills for this account decrease the pending bill counter. When the script finishes and no pending bills are present or a time-out for bills has occurred, whatever comes first, the account is closed, and a bill is requested.

- A **Billing Component** takes care of the mapping of the services and actions, taken place as recorded by the accounting component to a bill expressed in a certain currency. The bill component enables the agent to do billing based on accounting data. The bill consists of following three entries: description, quantity and price. A bill is automatically requested when an account is closed.

- A **Debugging Component** represents the debugging facilities of the agent. Debugging is one of the main aspects of JIAC to encourage secure, reliable applications. This component can be added on demand during runtime. Such a demand is initiated by a debugging-tool which requests the agent to add the debugging component to allow introspection of the agent by the debugger. That is done by the component by taking control of the backbone and forwarding certain vital messages to the debugger for further processing or displaying.

- A **Knowledge Base** contains the certain, selected facts of the environment the agent acts in. The knowledge base of the agent contains the global data. That is, for example, information about scripts or registered agents. Each component can access this knowledge with simple method calls because the knowledge base is a horizontal component, which means each component has a reference to it. That is an exception to the rule that each service of a component is accessible by speech acts.

- A **Runtime Component** controls the running scripts. Therefore, it can restrict the total number of simultaneously running scripts to a certain limit (the default is 50), and it monitors the execution time. In addition, this component creates unique names for each running script. This is necessary to allow the parallel execution of different instances of the same script class. Since JIAC components are named and scripts are components, too, each script has a name. This is also the point where the instance of a script, selected by the planner, is created. The speech act is the trigger of the script, and it contains all necessary data for the script execution. In order to allow the script to be called as a subscript of another script it is necessary to implement a second constructor, which takes all arguments of the triggering speech act. The script itself is started with a so called *addComponent*-call (via speech act). Vice versa, a speech act *'script-exited'* is sent to this component if the script has halted. The script is then removed from the internal data structures of this component. Also the exports of the script are added or removed, respectively, from the ServiceBroker. This is a good example of adding/removal of components during runtime. The finishing of one script enables the runtime component to start another script, which is in its pending queue when the script limit is reached. This is a convenient way to specify agent behaviour in Java. Each agent can hold a certain amount of scripts in its runtime component.

Scripts, by following a certain structure, are the commands the agent acts on, and they define its behaviour. By including more than one script, the agent can change its behaviour dynamically, by just selecting another script of its stock for execution, given a certain stimulus. This makes it possible to personalise and customise the agents in a very fine granularity, which is essential for heterogeneous environments like telecommunication networks.

```
<Script> :: <initScriptElement>
              {<ScriptElement>}*
<ScriptElement> ::= <ScriptElementName>:
              <BLOCK>
              RETURN <ScriptElementName>
<BLOCK> ::= <JavaCode>
  <initScriptElement> ::= <ScriptElement>
```

8 The Tools of JIAC

Tools for the support of AOT are the topic of this chapter. In contrary to agent architectures, where certain concepts are build in, suitable tools for AOT providing only partial solutions for certain aspects of the process of agentorented programming. The granularity of these partial solutions is more dispersed than in agent platforms e.g. in the support of distributed applications.

8.1 An agentoriented debugger

8.1.1 Requirements

For the developer and administrators of agent systems the following two question needs to be answered:
- **Monitoring:** Where are the agents and which communication relations exists
- **Debugging:** How can certain behaviour of the agents be explained that is how and why does the agent something

A monitor is suitable for the administration of agent systems without debugging functionality. However, a debugger is the integration of a monitor and a controller.

Visualisation
The basic requirement for a reasonable good work with a debugger is the presence of dedicated visualisation mechanisms for agent systems or parts of it. The following criteria's have to be taken in account:

- **Connection:** The user can connect to agent platforms or single agents he is interested in. After the connection is established the state of the system or the agent must be visualised to allow further actions of the user.
- **Distribution:** An agent systems is defined by the places agents can be and the actual place an agent is located on. This relationship must be made explicit.
- **Views:** The user should have the possibility to inspect certain interesting parts of the system. These parts are the capabilities of agents, the infrastructure and states of the agent system. Since these parts tend to be complicated it is necessary to have zoom functionality.
- **Cooperation:** Cooperation is realised through communication between agents. For the monitoring part the participants are of interest and for the debugging part the actual content of the message is important.

Debugging
The analysis of agent systems with the help of a debugger needs not only a visualisation but also a support of the manipulation of the system and the controlling of the whole system. Since the behaviour of an agent is deduced from its communication and its actions both parts have to integrated into the debugger equally.
The following requirements are inferred from the above statement:

- **Program-sequence:** During the execution of a program- the application of capabilities- it should be possible to make breakpoints and step through the code. Since capabilities can be loaded on demand it has to be possible to set a breakpoint newly invoked capabilities.
- **Messages:** The sending and receiving of message should be trace so that the system can be stopped in case of certain messages. Therefore the user has the possibility to get an alert when certain messages arrive and to inspect the systems afterwards.
- **Making tests:** To change the actual sequence of actions within the system to test the stability or robustness is also nice to have. That could be done by manipulating, generating or disposing messages and by the manipulation of the contents of the knowledge base of agents.

The following requirements are introduced by JIAC itself:
- Then a marketplace is inspected the agents on them should be made visible.
- The inspection of an agent by the debugger should show the components according to its type. These types are:
  - Agentcomponents
  - Capabilities – scripts – not used at the moment
  - Used capabilities

The different parts of the system have the following state changes which must be visualised:
- Marketplace:
  - Creation and elimination of agents
- Agents:
  - Creation and elimination of components
  - Location

- • Used capabilities
- • Mobility
- • Sending and receiving of speech acts
- • Scripts:
  - • State: possible/running/pending
- • Components:
  - • Description

Following the basic idea of JIAC that an agent consist of a number of different components, optimised for its particular domain, the debugging capability of an agent is realized by a dedicated Debug-Component. That allows to dynamically ad or remove the debugging functionality from the agent since it is only one more capability.

8.1.2 The views of the agentsystems

Views are realised by the debug component and the available information of the agent system.

Animation-View

The Animation-View shows all currently inspected marketplace and the agents on them in a 3 dimensional way. Additionally the messages and the migration of agents is shown as an 3D-animation.

The animation speed and the general appearance can be adjusted in a generous way. It is possible to navigate through the agent where new views are spawned according to certain aspects of the agent.

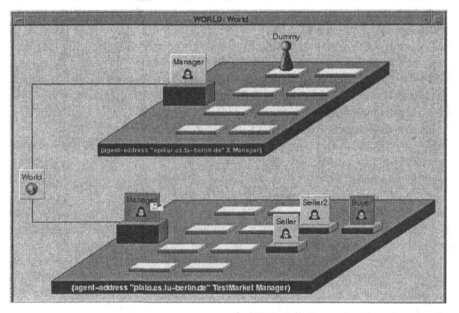

Figure 4: the Animation-View

Agent-View

The Agent-View is the standard view for agents. It shows all the components and the possible/running/pending scripts the agent consists of . Also the received speech acts are shown. The current location and the current status of the agent is displayed.

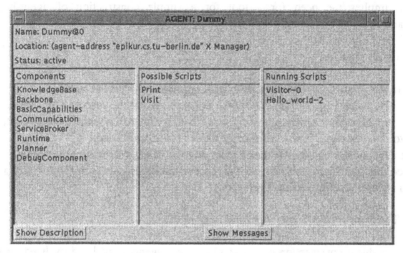

Figure 5: the Agent-View

Debug-Script-View

The debugging of scripts is only possible if the script is currently running. In the Debug-Script-View the possible states of a script are shown as a graph. The call of subscripts is highlighted.

The script elements can be processed step by step or breakpoints can be set.

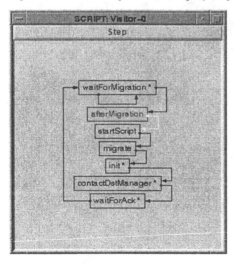

Figure 6: the Debug-Script-View

8.2 An agent and marketplace constructor tool

The programming of complex agent API's needs much time for acquaintance and construction of a single program.

That implies the need for a tool for a simple construction of agent-based applications.

8.2.1 Requirements

A graphical construction of marketplaces, agents and scripts is needed.

Also the agents have to be given their capabilities in the sense of JIAC these are scripts. It is needed to generate Java-Source-Code, for later fine tuning, and to compile this into Java-classes. These classes can than be started and tested.

The constructor should offer the connection to tools like the monitor/debugger.

Since the agents are only as good as there capabilities it is required to manage a large database of scripts for the agents.

8.2.2 The Constructor-View

The graphical construction of marketplaces and agents is realized by heeding the possibilities offered by the Java-Beans.

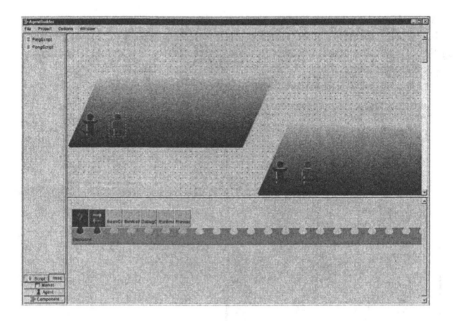

Figure 7: The Constructor View

8.3 A marketplace administration tool

Electronic marketplaces require a comfortable administration of the different service provider agents on them. The complexity of the demands is a result of the number of different distributed marketplaces.

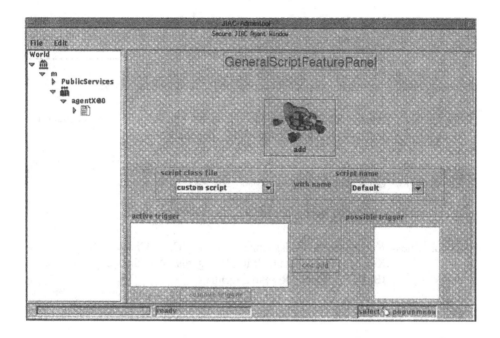

Figure 8: JIAC-Admintool

The aspect of the dynamic exchange of agents on markets is taken care of by JIAC itself. Onle the appropriate graphical interfaces have to be programmed. More effort needs the maintenance of the agent and script libraries.

9 Conclusions and Outlook

JIAC is a reliable and secure environment for mobile agents under Java that is especially suitable for applications in electronic commerce and telecommunication services. For these purposes, JIAC comes with a complete set of classes for agents and marketplaces. JIAC agents offer graphical tools for debugging, building of applications and administration.

Furthermore, JIAC is designed from the bottom up following a very modern component approach that makes it possible to build agents and marketplaces by reusing tested components which ends up in more reliable and more cost efficient systems. This plug-and-play approach makes JIAC a very promising candidate for further agent oriented software projects and future research. Since JIAC is an ongoing project with several applications, it is constantly extended by capabilities and functionality.

References

[Hohl, 1995]   Hohl, Fritz, "Konzeption eines einfachen Agentensystems und Implementation eines Prototyps", Universität Stuttgart, Fakultät Informatik, Diplomarbeit Nr. 1267 (1995)

[Odyssey, 1997] "Introduction to the Odyssey API", General Magic, 1997

[Gray, 1995]   Robert S. Gray, "Agent Tcl, alpha release 1.1", Department of computer science, Dartmouth College, 1995

[Voyager, 1997] "Voyager – Core Technology User Guide Version 1.00", ObjectSpace, 1997

[Aglets, 1997]   "Aglets Workbench", IBM Japan, 1997

[Harrison, Chess, Kershenbaum] Harrison, Colin G., Chess, David M., Kershenbaum, Aaron, "Mobile Agents: Are they a good idea?", IBM T. J. Watson Research Center

[Java]   http://java.sun.com

[KQML]   http://www.cs.umbc.edu/kqml/papers

[Gamma, Helm, Johnson, Vlissides, 1994] Gamma, Erich; Helm, Richard; Johnson, Ralph; Vlissides, John: "Design Patterns", Addison Wesley, 1994

[Chauhan, 1997] Chauhan, Deepika, "JAFMAS: A Java-based Agent Framework for Multiagent Systems Developement and Implementation", ECECS Department, University of Cincinatti, 1997

[AgentBuilder, 1998]   Reticular Systems, "AgentBuilder White Paper", 1998

Grasshopper - A Mobile Agent Platform for Active Telecommunication Networks

C. Bäumer, T. Magedanz
IKV++ GmbH
Kurfürstendamm 173-174, D - 10707 Berlin, Germany
Email: [baeumer | magedanz] @ikv.de

Abstract. This paper presents an overview of the Grasshopper agent platform, developed by GMD FOKUS and IKV++ GmbH, which provides a powerful middleware for the implementation of agent-based telecommunication services, such as advanced network management, intelligent network, and mobile communications applications. The main target is to enable the dynamic deployment of management and control services inside the network nodes, thereby making the telecommunication networks active. Besides a general overview of the Grasshopper platform features and its usage within some European ACTS research and development projects, we will briefly illustrate the usage of Grasshopper for the implementation of an Active Broadband Intelligent Network environment. This network environment is currently realised within the ACTS project MARINE (Mobile Agent-based Intelligent Network Environment).

Introduction

A new killer application for agent technology in the context of (tele)communications is *"Active Networking"*. This new buzzword describes the general vision of (tele)communication network evolution, where the network nodes become active because they take part in the computation of applications and provision of customized services. The basic idea of such Active Networks [1][2] is the movement of service code, which has been traditionally placed outside the transport network, directly to the network's switching nodes. Furthermore, this movement of service code should be possible in a highly dynamic manner. This allows the automated, flexible, and customized provision of services in a highly distributed way, thereby enabling better service performance and optimized control and management of transport capabilities. Two enabling technologies are key in this context: *programmable switches*, which provide flexibility in the design of connectivity control applications [3], and *mobile code systems/mobile agent platforms*, which enable the dynamic downloading and movement of service code to specific network nodes [4].

This paper introduces the Grasshopper agent platform as an enabling technology for the short term implementation of active telecommunication networks. Today Grasshopper is the agent platform of choice in multiple international ressearch projects within the European CLIMATE *(Cluster for Intelligent Mobile Agents for Telecommunication Environments)* initiative. CLIMATE explores the usage of agent-based middleware in particular telecommunication domains, such as service control in fixed and mobile networks, telecommunications management, multimedia appl i-cations, etc.

Based on previous research activities and related publications [5][6] this paper illustrates the usage of Grasshopper for the implementation of an active *Broadband Intelligent Network (B-IN)* environment. This active B-IN environment enables the dynamic deployment and distribution of MA-based services onto enhanced broadband switching equipment and service nodes.

In the following section a detailed overview of the capabilities of the Grasshopper agent platform is given. This section lists also some important telecommunications research projects, which are based on Grasshopper. In section 3, we illustrate how Grasshopper is used for realising an active B-IN environment, which is currently implemented within the ACTS MARINE project. Section 4 concludes this paper.

1 Grasshopper – The Agent Platform

Grasshopper [7], which has been developed by GMD FOKUS and IKV++ GmbH, is an agent development and runtime platform that is built on top of a distributed processing environment. It is based on Java JDK 1.2 and written in Java. Furthe r-more, Grasshopper is conformant to the major agent standards addressing interoperability of agents, namely the *OMG Mobile Agent System Interoperability Facility (MASIF)* [8] and the *Foundation of Intelligent Physical Agents (FIPA)* `97 specifications [9]. Grasshopper is in principle a MASIF conformant mobile agent platform, which has been enhanced recently with a FIPA add on, in order to give the application developer total flexibility. This evolution of the platform is witnessing the fact, that the traditional separation of mobile agents and intelligent agents is going to fade away smoothly, as the corresponding standards bodies, i.e., OMG (namely the new Agent SIG [10]) and FIPA, are aiming to develop compatible standards. So Grasshopper enables its users to develop a broad range of agents, ranging from small simple mobile agents able to roam within the network up to static multi agent systems talking via an Agent Communication Language (ACL) for distributed problem solving.

In the following we describe the basic features of Grasshopper enabling the deve-lopment mobile agents as well as the new FIPA Add-On becoming available in 1999.

1.1 Basic Grasshopper Concepts and Functionality

In principle, Grasshopper realises a *Distributed Agent Environment* (DAE). The DAE is composed of regions, places, agencies and different types of agents. Fig 1 depicts an abstract view of these entities.

Two types of agents are distinguished in Grasshopper: mobile agents and station-ary agents. The actual runtime environment for both mobile and stationary agents is an *agency*: on each host at least one agency has to run to support the execution of agents. A Grasshopper agency consists of two parts: the core agency and one or more places. Core Agencies represent the minimal functionality required by an agency in order to support the execution of agents. The following services are provided by a Grasshopper core agency:

- *Communication Service*

  This service is responsible for all remote interactions that take place between the distributed components of Grasshopper, such as location-transparent inter-agent communication, agent transport, and the localization of agents by means of the region registry. All interactions can be performed via CORBA IIOP, Java RMI, or plain socket connections. Optionally, RMI and plain socket connections can be protected by means of the Secure Socket Layer (SSL) which is the de-facto standard Internet security protocol. The communication service supports synchronous and asynchronous communication, multicast communication, as well as dynamic method invocation. As an alternative to the communication service, Grasshopper can use its OMG MASIF-compliant CORBA interfaces for remote interactions. For this purpose, each agency provides the interface *MAFAgentSystem*, and the region registries provide the interface *MAFFinder* [9].

- *Registration Service*

  Each agency must be able to know about all agents and places currently hosted, on the one hand for external management purposes and on the other hand in order to deliver information about registered entities to hosted agents. Furthermore, the registration service of each agency is connected to the region registry which maintains information of agents, agencies and places in the scope of a whole region.

- *Management Service*

  The management services allow the monitoring and control of agents and places of an agency by (human) users. It is possible, among others, to create, remove, suspend and resume agents, services, and places, in order to get information about specific agents and services, to list all agents residing in a specific place, and to list all places of an agency.

- *Security Service*

  Grasshopper supports two security mechanisms: external and internal security.

  - External security protects remote interactions between the distributed Grasshopper components, i.e. between agencies and region registries. For this purpose, X.509 certificates and the Secure Socket Layer (SSL) are used. SSL is an industry standard protocol that makes substantial use of both symmetric and asymmetric cryptography. By using SSL, confidentiality, data integrity, and mutual authentication of both communication partners can be achieved.

  - Internal security protects agency resources from unauthorized access by agents. Besides, it is used to protect agents from each other. This is achieved by authenticating and authorizing the user on whose behalf an agent is executed. Due to the authentication/authorization results, access control policies are activated. The internal security capabilities of Grasshopper are mainly based on JDK security mechanisms.

- *Persistence Service*

  The Grasshopper persistence service enables the storage of agents and places (the internal information maintained inside these components) on a persistent medium. This way, it is possible to recover agents or places when needed, e.g. when an agency is restarted after a system crash.

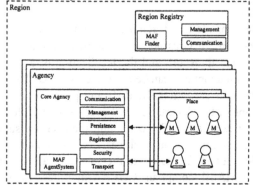

Fig. 1. The Grasshopper Distributed Agent Environment

A place provides a logical grouping of functionality inside of an agency. The region concept facilitates the management of the distributed components (agencies, places, and agents) in the Grasshopper environment. Agencies as well as their places can be associated with a specific region by registering them within the accompanying region registry. All agents that are currently hosted by those agencies will also be automatically registered by the region registry. If an agent moves to another location, the corresponding registry information is automatically updated.

The functionality of Grasshopper is provided on the one hand by the platform itself, i.e. by *core agencies* and *region registries*, and on the other hand by *agents* that are running within the agencies, in this way enhancing the platform's functionality. The following possibilities regarding the access to the Grasshopper functionality must be distinguished:

- Agents can access the functionality of the *local agency*, i.e. the agency in which they are currently running, by invoking the methods of their super classes `Service`, `StationaryAgent`, and `MobileAgent`, respectively. These super classes are provided by the platform in order to build the bridge between individual agents and agencies, and each agent has to be derived from one of the classes `StationaryAgent` or `MobileAgent`.
- Agents as well as other DAE or non-DAE components, such as user applications, are able to access the functionality of *remote agencies* and *region registries*. For this purpose, each agency and region registry offers an external interface which can be accessed via the Grasshopper communication service.
- Agencies and region registries may optionally be accessed by means of the MASIF-compliant interfaces `MAFAgentSystem` and `MAFFinder`.

In the context of Grasshopper, each agent is regarded as a *service*, i.e. as a software component that offers functionality to other entities within the DAE. Each agent/service can be subdivided into a common and an individual part. The common (or core) part is represented by classes that are part of the Grasshopper platform, namely the classes `Service`, `MobileAgent`, and `StationaryAgent`, whereas the individual part has to be implemented by the agent programmer.

A Grasshopper agent consists of one or more Java classes. One of these classes builds the actual core of the agent and is referred to as *agent class*. Among others, this class has to implement the method `live` which specifies the actual task of the agent. The agent class must be derived either from the class `StationaryAgent` or from the class `MobileAgent` which in turn inherits from the common super class `Service`. The methods of these classes represent the essential interfaces between agents and their environment. The following two ways of method usage have to be distinguished:

- One part of the super class methods of an agent enable the access to the local core agency. For example, an agent may invoke the method `listMobileAgents()`, which it inherits from its super class `Service`, in order to retrieve a list of all other agents that are currently residing in the same agency.
- The remaining super class methods are defined to access individual agents. These methods are usually invoked by *other* agents or agencies via the *communication service* of Grasshopper. For instance, any agent may call the method `getState()` of *another* agent in order to retrieve information about the other agent's actual state. Note that this way of access is not performed directly on an agent instance, but instead on an agent's *proxy* object.

1.2 The Grasshopper FIPA Add-On

Due to the increasing acceptance of the FIPA standards [9] and the resulting increased demand for FIPA conformant agent platforms, Grasshopper has been extended by a corresponding package, referred to as *"FIPA Add On"* in order to allow agents to communicate via the FIPA Agent Communication Laguage (ACL) [11].

Part 1 of the FIPA standard [12] proposes the concept of an Agent Platform (AP) comprising of three basic services being offered. These services are namely the Agent Management System (AMS), the Directory Facilitator (DF) and the Agent Communication Channel (ACC). Agents are considered residing on a home agent platform (HAP) if they are registered on the HAPs AMS. Agents may offer their services to other agents and make their services searchable in a yellow pages manner by the DF if they register on the DF. Registration on a DF is voluntary while registering on the AMS is mandatory for being an a AP. Finally the ACC is enabling agent communication (AC) between agents on a platform and between platforms by offering a message forwarding service called *forward*. Reachability between platforms is gained by making the forward service available by IIOP.

Grasshopper provides the main components of a FIPA complaint platform as depicted in Fig. 2 in form of corresponding stationary Grasshopper agents, namely an

Agent Management System (AMS), a Directory Facilitator (DF), and an Agent Communication Channel (ACC).[1]

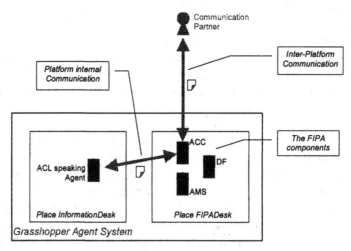

Fig 2   Realisation of the FIPA Platform Components on Top of Grasshopper

All these FIPA platform components are based on the class *FIPAAgent*, which extends Grasshoppers class *StationaryAgent*. Agent developers have also to use this class as basis for their agent application. The class FIPAAgent offers two methods: *send()* and *message()*. With them, the ACL-message exchange over the ACC is realised. The method send() requires a parameter of the type ACLMessage and server for dispatching of messages, whereas the message method serves for retrieval of messages.

According to the FIPA standard the Grasshopper AMS is responsible for the managing of operation of the agent platform. This comprises the management of the agent and the agent state. All these features are already provided by Grasshopper. Thus, the *AMSAgent* provides a FIPA compliant interface to the Grasshopper agent management facilities.

The DF realises a ´yellow page´ in a FIPA compliant platform. In Grasshopper this functionality is already realised by the *Region Registry*. Thus, the DF is realised as a wrapper of these features.

The ACC component is also implemented as a Grasshopper agent. Inheriting from the class FIPAAgent it will provide the *message* and *send* methods. The ACC agent will be responsible for the communication of FIPA agents[2] residing on different agent systems. For this, the ACC agent has to be available on each agent system belonging to an FIPA conformant environment. To support IIOP which is necessary for interoperability the ACC agent has to be provided with a CORBA interface. Ac-

---

[1] To support platform interoperability as required by FIPA the ACC has to support IIOP. The current Grasshopper version 1.2 supports this protocol since this Java JDK contains a complete ORB including IIOP.

cording to the particular CORBA implementation the ACC agent inherits from a corresponding provided superclass. Then, as a CORBA object the ACC agent's methods can be invoked via IIOP.

The Grasshopper FIPA Add-On implementation supports the new standard language for the World Wide Web, i.e., the eXtensible Markup Language (XML) as well the FIPA ACL as of FIPA 97 Specification (part 2). An appropriate parser for FIPA ACL has been implemented and is provided to the FIPA agents, i.e., the FIPA platform components ACC, DF, AMS as well as to the applications. Using the public available XML engine from IBM makes it easy to parse the incoming XML message. With XML, appropriate ontologies can be specified for different application domains. The communication between the FIPA agents take place with the default communication language, either FIPA ACL or XML ACL, which is selectable at the starting of the FIPA platform. The real message content e.g., message payload can be encoded in FIPA SL1 or also in XML without any extra user intervention. To support any other (perhaps proprietary) content language, it is left to the users, to develop their specific own parser implementation.

To implement an agent which intends to act as a FIPA agent, the agent class has to inherit from the class FIPAAgent by means of writing a Java class extending FIPA.FIPAAgent. The class FIPAAgent is itself extending the class de.ikv.grasshopper.agency.StationaryAgent, thus being a regular grasshopper stationary agent. Apart from the regular Grasshopper agent methods, which have to be implemented for each Grasshopper agent such as the *live* method, the agent developers have to implement (overwrite) the *message* method, whereas the *send* method can be simply used.

1.3 Usage of Grasshopper for Advanced Telecommunications

Grasshopper can be used in many different application contexts. Historically, the telecommunications domain is the most prominent application areas but not the only one. Grasshopper is used today in several European research and development projects, such as ANIMA, EURESCOM P815, CAMELEON, FACTS, MARINE, MARINER and MIAMI. Most of these projects belong to CLIMATE [13], the Cluster for Intelligent Mobile Agents in Telecommunication Environments, which is part of the European Research Programme on Advanced Communication Technologies and Services (ACTS). A short description of these projects and the usage of Grasshopper is given below.

1.3.1 ANIMA - Architecture Neutral Intelligent Mobile Agents

The objective of the ESPRIT ANIMA project [14] is to develop and validate agent based solutions to support the dynamic deployment of commercial applications on easy-to-use terminals such as Windows CE based screenphones. Therefore ANIMA applications need to be able to run on deeply embedded systems which have extremely tight memory space, a low-bandwidth network connection and a limited physical user interface (screen area and input medium). Therefore, the ANIMA core will need to be adapted and shrunk to a small memory footprint, and the network bandwidth will need to be reduced to a minimum.

Grasshopper is used as basis of the ANIMA platform extended with various facilities as required by the deployment in a commercial oriented end user environment. Mainly, these extensions regard to the support of Windows CE based devices as well as to the areas of security, platform and service management.

1.3.2 CAMELEON – Communication Agents for Mobility Enhancements in a Logical Environment of Open Networks

The CAMELEON project [15] aims for an agent-based implementation of the Virtual Home Environment concept, considered key for 3rd generation mobile communication systems (eg., UMTS). Key is universal adapted service access for roaming users in mobile environments.

In this project Grasshopper forms the basis for the realisation of enhanced customer service control applications, as well as agent-based service access, delivery, customisation and mobility management. An important aspect is to make Grasshopper also available for Windows CE based terminals, considered important devices in the UMTS context.

1.3.3 EURESCOM Project 815 – Communication Management Process Integration Using Software Agents

The EURESCOM P815 project [16] main goal is to analyse, design and develop a prototype of an agent-based workflow management system that can deployed for the automation of inter-domain Telecommunication Business Processes. Additionally, the project will develop, testify and validate a set of appropriate inter-domain business process scenarios. The selected scenarios are the International Leased Line Provision (ILLP) scenario and Intelligent Networks scenario. For every scenario, a set of intelligent agents providing the individual functional steps of the business processes will be developed and tested.

The FIPA compliant version of the Grasshopper Agent Platform will be the underlying development platform for the development of the workflow management system, as well as, the development of the business process intelligent agents. Additionally, integration of the agent-based workflow management system and interoperation of Grasshopper platform with other FIPA compliant agent platforms will be performed.

1.3.4 FACTS - FIPA Agent Communication Technologies and Services

The goal of the FACTS (FIPA Agent Communication Technologies and Services) project [17] will be to validate the work of FIPA and other standards groups by constructing a number of demonstrator systems based on proposed standards. Note that the focus of the project is in the interaction between differently-implemented agents, and not on agent technologies or on the application domains that will be used as a vehicle for the project.

Grasshopper is used by GMD FOKUS and Alcatel Belgium as the development and test platform for an Agent-based VPN application. The developments (done in parallel by the two partners) will mainly utilise the FIPA ACL extensions and XML/RDF support on top of Grasshopper.

1.3.5 MARINE – Mobile Agent Environment for Intelligent Networks

MARINE [18] introduces agent platforms into a broadband IN architecture, enabling the provision of agent-based IN services directly at the distributed IN switches, which by the introduction of agent platforms become IN Service Nodes. This enables decentralised service provision, decreasing signalling network load.

Grasshopper forms the technical basis for this project. In addition to the development of Grasshopper based broadband IN services and making use of wrapped signalling interfaces (B-INAP), an important aspect is the development of a Grasshopper-based service management system for Grasshopper-based service environments.

Section 3 looks in more detail on this project.

1.3.6 MARINER – Multi-Agent Architecture for Distributed-IN Load Control and Overload Protection

MARINER [19] is introducing agent technology for traffic control in existing Intelligent Networks, where agents located at IN switches (SSPs) and central service control nodes interact for strategic decision-making. The focal point is on Agent Communication based on the FIPA specifications.

For this project Grasshopper is enhanced with a FIPA Add On (providing FIPA conformant ACL-Interfaces) and used for the implementation of appropriate load balancing agents which exchange Signaling System No. 7 load and node load information.

1.3.7 MIAMI – Mobile Intelligent Agents for Managing the Information Infrastructure

MIAMI [20] aims to develop mobile agent solutions for the management of the Open European Information Infrastructure and to create a unified mobile intelligent agent framework which supports those solutions in an adequate way. The MIAMI project on the one hand uses Grasshopper as a basis to develop new or enhance existing features of the Grasshopper platform. The main areas are security, resource management, logging services, agent communication transport services, and high level agent communication services. On the other hand, the MIAMI project uses Grasshopper for the deployment of mobile intelligent agents developed in the project. Those agents are used for purposes like the automated establishing and managing of virtual enterprise together with workflow management supporting agents. They are also used for a variety of network management purposes such as fault, performance, and configuration management.

2 Using Grasshopper for Implementing an Active B-IN

2.1 Intelligent Network Limitations and Evolution

Among the different solutions aimed at providing advanced telecommunication services, the Intelligent Network (IN) represents at the end of this century the most prominent architecture [21]. The reason for this is that the IN provides a uniform and extensible service platform which should enable the rapid introduction of customized telecommunication services across different bearer networks, such as PSTNs (Public Switched Telephone Networks), ISDNs (Integrated Digital Services Networks) and Broadband ISDN (B-ISDN).

The main architectural principle of the IN is the separation of service switching and service control, with a reduced number of centralized nodes (SCPs, Service Control Points) hosting service logic and data for controlling via a dedicated IN Application Protocol (INAP) on top of the Signalling System No. 7 (SS7) network a high number of distributed specialized switches (SSPs, Service Switching Points). In addition, special assistant devices (IPs, Intelligent Peripherals) provide additional capabilities for advanced user interactions, which for cost reasons can not be accommodated in all switches. IN services are deployed and manged via a Service Management System (SMS), which obtains the services from a Service Creation Environment (SCE). Users gain access to the services via their terminal equipment. For more details look at [21].

The increasing competition between network operators requires fast responses to users' needs. Service deployment times represent a key success factor for operators. In order to accomplish this objective, revisions of equipment available in the network, essentially concentrated on software changes, become a crucial requirement for the success of a technical solution. The quite centralized approach adopted in the IN architecture is a consequence for supporting this requirement, since services have to be introduced only in the centralised control nodes and not in all switches. However, in face of an increasing number of IN services, the centralized service control nodes (and the signalling network) become performance bottlenecks.

Since IN evolution takes into account also recent progress of the IT domain, new opportunities exist to tackle these threads. The availability of new software technologies such as *DOT (Distributed Object Technology)* and *MAT (Mobile Agent Technology)* allows to evolve the current IN architecture, looking for systems where intelligence can be distributed where and when needed, while maintaining compatibility with current centralized architectures. This solution allows to design dynamic / active IN architectures, where enhanced switching systems can take over the most adequate role from time to time, depending on the software capabilities they host in a given time frame. In the following we briefly describe the approach taken by the ACTS project *MARINE (Mobile Agent enviRonments in Intelligent Networks)* project. MARINE [18] which uses the Grasshopper platform for implementing an active B-IN.

2.2 The MARINE Project Implementing an Active B-IN

In the scope of the MARINE project, advanced broadband services, such as broadband video telephony and video on demand, are realized by means of mobile service agents within an active B-IN environment. These service agents can be provided time-dependent, i.e., installed for a limited time duration, and location dependent, i.e., installed at dedicated switches or even in specific end user systems, and thereby relax the load of service control and service management systems and the corresponding signalling and data networks.

However, MARINE adopts an evolutionary approach for the IN, taking into account an interworking between the new Grasshopper-based service environment, comprising multiple switching nodes enabled to run services locally (i.e., broadband Service Switching and Control Points, B-SSCPs) and central service execution nodes (i.e., broadband SCPs, B-SCPs), and the traditional broadband IN environment based on B-SSPs and B-SCPs. This means that it should be possible for a new open switches to access transparently remote B-SCP based services if necessary. On the other hand, a traditional B-SSP should be able to access new MA-based services in a centralized service execution node. The overall scenario is depicted in Fig 3.

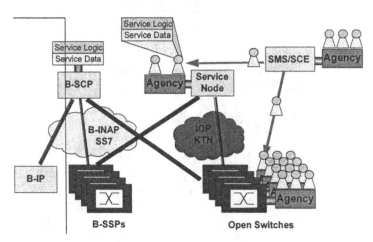

Fig. 3. The MARINE Reference Architecture

The MARINE DAE is structured as follows. By means of the region concept, agencies belonging to a single network or service provider are grouped; i.e., each network operator has its own region. The place concept is used to separate IN related capabilities inside a single agency. For example, an agency within the B-SSCP hosts a place for SSF related agents as well as a co-located SCF/SDF place for the service agents (see also Fig 4). [3]

Furthermore, two kinds of agents are required in addition to the core agency services: Firstly, specific service agents related to the multimedia service environment/infrastructure (e.g., basic bearer connectivity, special resource access, interworking/gateway services, etc.) have to be provided, which are not mobile, since they are related to specific locations (e.g., a switch, special resources, etc.). Secondly, the mobile agents implementing the actual B-IN service logic have to be provided.Looking briefly at the first class of environment agents, the B-SSP is enhanced by a Grasshopper agency, which provides at least two places, namely:

- a B-SSF place which provides a connectivity service to the broadband Switching State Manager (B-SSM) and a service trigger agent responsible for dispatching service requests to (local or remote) service agents, and
- a B-SCDF (SCF/SDF) place hosting all the local service agents and a housekeeper agent.

Optionally, a B-SRF place provides service capabilities for the access to dedicated special resources, such as speech synthesis, video codecs, etc. Thus, this place hosts service agents providing customized announcements, video previews, etc.

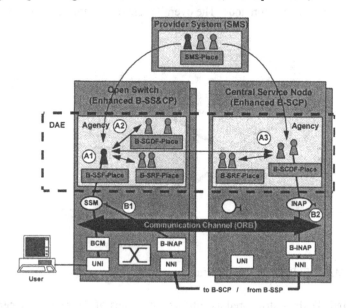

Fig. 4. Structure of the MARINE Distributed Agent Environment

---

[3] Note that in this context we adopt the notion of IN functional entities, i.e., Service Switching Function (SSF), Service Control Function (SCF), Service Data Function (SDF), and Specialised Resource Function (SRF).

As there is the need for centralized service provision (e.g., for mobility services), the B-SCP will be enhanced by the aforementioned B-SCDF (SCF/SDF) place, featuring besides inter-ORB communication also a B-INAP stack and a corresponding B-INAP/CORBA gateway for allowing access from traditional B-SSPs. Furthermore, the enhanced B-SCP may accommodate an SRF place.

Finally, an SMS place exists in the provider system, which serves as service agent repository and in addition provides appropriate management services for the MARINE environment in order to control and monitor the mobile service agents. Service agents are always starting their itinerary from this agency. This itinerary may be pre-configured, based on the service type and service user locality.

Fig 4 depicts the MARINE Reference Configuration in a more detailed way. Via the B-SSM, the B-SS&CP is able to invoke the CORBA-based local and remote service agents (indicated as A1 - determination of local or remote service agents, A2 - invocation of a local service agent, and A3 - invocation of a remote service agent) as well as traditional B-INAP-based services (B1 in Fig 4). Note that this B-INAP request is sent to a traditional B-SCP. As also indicated in Fig 4, the enhanced B-SCP may be accessed via CORBA and agent communication mechanisms as well as through B-INAP requests coming from traditional B-SSPs (B2).

For a more detailed technical description of the MARINE architecture readers are referred to [22].

3 Conclusions

This paper has described the Grasshopper agent platform representing the enabling technology for many European agent projects in the context of telecommunications. Grasshopper provides a high degree of flexibility for software developers to realise their ideas. Therefore Grasshopper is a strong candidate for the implementation of future agent based applications including active networking environments as shown in the context of the MARINE project. A comparison of Grasshopper with other MA platforms, such as Aglets Workbench, Concordia, Odyssey, and Voyager is given in [4].

Future application domains for Grasshopper will be the ongoing integration of voice and data networks, unified messaging, and active networking. Furthermore, a major emphasis will be placed on using Grasshopper for electronic commerce applications, e.g., financial services, and the implementation of distributed internet search engines, where the integration of existing web information is a pivotal importance. Corresponding projects are underway in the 5th framework of the European Community known as Information Society Technologies (IST).

References

1. L. Tennenhouse, et.al. (1997). 'A Survey of Active Network Research', IEEE Communic a-tions Magazine, pp. 80-85, Vol. 35, No. 1, January 1997
2. MIT Active Networks homepage: http://www.tns.lcs.mit.edu/activeware

3. P1520 - Proposed IEEE Standard for Application Programming Interfaces for Networks: http://www.iss.nus.sg/IEEEPIN/

4. M. K. Perdikeas, F. G. Chatzipapadopoulos, I. S. Venieris, G. Marino: "Mobile Agent Standards and Available Platforms", Computer Networks Journal, Special Issue on "Mobile Agents in Intelligent Networks and Mobile Communication Systems", ELSEVIER Publisher, Netherlands, vol. 31, issue 10 (1999)

5. T. Magedanz, R. Popescu-Zeletin (1996b). 'Towards Intelligence on Demand - On the Impacts of Intelligent Agents on IN', Proceedings of 4th International Conference on Intelligent Networks (ICIN), Bordeaux, France, December (1996) 30-35

6. M. Breugst, T. Magedanz: "Mobile Agents - Enabling Technology for Active Intelligent Networks", IEEE Network Magazine, Vol. 12, No. 3, Special Issue on Active and Programmable Networks (1998) 53-60

7. IKV++ GmbH – Grasshopper homepage: http://www.ikv.de/products/grasshopper

8. OMG (1995), Common Facilities RFP3, Request for Proposal OMG TC Document 95-11-3, Nov. 1995, http://www.omg.org/; MASIF specification is available through http://ftp.omg.org/pub/docs/orbos/97- 10-05.pdf

9. FIPA homepage: http://www.fipa.com

10. OMG Agent SIG homepage: http://www.objs.com/isig/agents.html

11. FIPA TC2: Agent Communication Language, April 1997

12. FIPA TC1: Agent Management, April 1997

13. CLIMATE homepage: http://www.fokus.gmd.de/research/cc/ima/climate/climate.html

14. ANIMA project homepage: http://anima.ibermatica.com/anima

15. CAMELEON project homepage: http://www.comnets.rwth-aachen.de/~cameleon

16. EURESCOM P815 project homepage: http://www.eurescom.de/public/projects/P800-series/P815/p815pf.htm

17. FACTS project homepage: http://www.labs.bt.com/profsoc/facts/

18. MARINE project homepage: http://www.italtel.it/drsc/marine/marine.htm

19. MARINER project homepage: http://www.teltec.dcu.ie/mariner

20. MIAMI project homepage: http://www.fokus.gmd.de/research/cc/ima/miami/

21. T. Magedanz, R. Popescu-Zeletin. Intelligent Networks - Basic Technology, Standards and Evolution', International Thomson Computer Press, ISBN: 1-85032-293-7, London (1996)

22. L. Faglia, T. Magedanz, A. Papadakis: "Introduction of DOT/MAT into a Broadband IN Architecture for Flexible Service Provision", H. Zuidweg et.al (Eds.), IS&N 99, LNCS 1597, ISBN: 3-540-65895-5, Springer-Verlag (1999) 469-481

Advanced Network Management Functionalities through the Use of Mobile Software Agents

Antonio Puliafito[1], Orazio Tomarchio[2]

[1] Dipartimento di Matematica, Università di Messina
C.da Papardo - Salita Sperone, 98166 Messina - Italy
E-mail: apulia@ingegneria.unime.it

[2] Istituto di Informatica e Telecomunicazioni, Università di Catania
Viale A. Doria 6, 95025 Catania - Italy
E-mail: tomarchio@iit.unict.it

Abstract. Computing power and storage space are more and more distributed over the network. Consequently, management strategies are becoming of crucial importance in order to guarantee an effective control of the system, both in terms of performance and reliability. Mobile agents represent a challenging approach to provide advanced network management functionalities, due to the possibility to easily implement a decentralized and active monitoring of the system. In this paper we discuss how to take advantage of this technology and describe a prototype implementation based on our mobile agent platform called MAP.

Keywords: network management, mobile agents, SNMP, Java.

1 Introduction

The growing complexity of computer networks requires the use of sophisticated management techniques. Monitoring, properly detecting and controlling the behaviour of system resources represent the main functionalities a network management system should provide.

In the approaches currently adopted by the main standard platforms, management data are stored in databases maintained on the elements to be managed. Management applications run on a central station, where data need to be transferred for providing high-level information, which are useful for the management of the system. According to this approach, the management of a network tends to require a considerable amount of bandwidth, particularly when some troubles in the network arise, and an immediate manager's action is required.

In this work, our purpose is exploiting new technologies based on mobile code, in order to extend such architectures by moving a portion of the *"intelligence"* to the nodes where data are resident. In this scenario, management agents are placed in each node: they monitor the state of the node, and can perform simple management functions on the local node. If agents find irregular operation conditions, a communication mechanism is triggered among the different agents, which activate the appropriate recovery strategies. For example, such strategies consist of system reconfigurations, of new routing strategies, etc. Many of the management decisions may be taken locally, thus avoiding to transfer large amounts of data from the remote nodes to a central station. Only high-level information will be transferred to the central station where the network manager operates. Besides, in the proposed infrastructure, agents are not statically resident in the remote devices. They can move from a node to another, can be replaced if necessary, can be equipped with functionalities that had not been envisaged at the time of installation. All of the management procedures do not need to be maintained in each agent; some of these procedures would migrate from a central node to the one where they will be executed, but only when they become really necessary. Such functionalities make the system obtained very flexible. Besides, the management application itself has been organized as a mobile agent; in this way the manager of the network no longer needs a fixed position.

A prototype of the network management system described before has been implemented by using MAP[3] (Mobile Agents Platform). It is a platform for the development and the management of mobile agents, which was completely developed by using Java at the University of Catania [8]. This platform gives the user all the basic tools needed for the creation of agent based applications. It enables us to create, run, suspend, resume, deactivate, reactivate local agents, to stop their execution, to make them communicate with each other and migrate. The use of Java (thanks to its independence from hw and sw architectures) enabled us to develop a platform able to operate in heterogeneous environments.

The rest of this paper is organized as follows: in Section 2 we present the basic concepts of traditional network management systems, and we outline their main limits. In Section 3 we discuss the main models of distributed management which have been created for overcoming such problems: we also outline the main strengths of an approach based on mobile code. In Section 4, after a short description of the platform MAP, the agents (expressly created) and their functionalities are described. Finally, in Section 5 we present our conclusions and the future hints for the research.

2 Traditional systems of network management

Current network management systems adopt a centralized paradigm according to which the management application periodically accesses the data collected by a set of software modules placed on the network devices, by using an appropriate protocol.

The *Simple Network Management Protocol (SNMP)* [3, 10, 11] proposed by the IETF has become the standard protocol of management for IP networks, while the *Common Management Information Protocol (CMIP)* [11] proposed by the ISO has been used only in the area of networks for telecommunications.

---

[3] MAP is available at http://sun195.iit.unict.it/MAP

Both architectures have a similar approach, and only differ for the way they operate. The basic components of such systems of network management are the following ones:

- one or more management stations (*Network Management Station or NMS*)
- some nodes (potentially many of them). A module called *agent*[4] runs on each node; it monitors and collects the data of the node.
- a *management protocol*, which is used for transferring the management information among the snmp-agents and the management stations.

Some management applications are executed in the NMS (on which the network administrators normally operate), for monitoring and controlling the network elements. They are devices such as hosts, routers, bridges, terminal servers, etc., which can be monitored and/or controlled by accessing their management information. The management information are considered as a collection of *managed objects*, stored in so-called *Management Information Base (MIB)* [7]. Some sets of related objects are defined in the MIB modules. Such modules are specified by using a subset of the standard notation OSI Abstract Syntax Notation One (ASN.1), called *Structure of Management Information (SMI)*. The snmp-agents of this infrastructure are entities providing such information with a standardized interface. The NMS interacts with these snmp-agents, acting as a client that (according to the indications of the network manager) asks the agents for information about the different network devices.

All the applications concerning the management are executed on the NMS. Snmp-agents have a very simple structure, and usually communicate only in response to the requests of variables contained in the MIB. They cannot perform management actions on their local data. The network management protocol provides the primitives for exchanging management informations among the snmp-agents and the management stations. The access to MIB variables by the NMS takes place at a low level. For example, the set of primitives of the SNMP is very simple, and provides three basic types of operations: the operators *set* and *get* for setting or reading the value of a variable, and the operator *getnext* for continuing to examine the subsequent variable in the MIB.

2.1 Limitations of the centralized paradigm of the SNMP

The centralized paradigm adopted by the SNMP is appropriate in several network management applications, but the quick expansion of networks has posed the problem of its scalability, as well as for any other centralized model. At the same time, the computational power of the network nodes to be managed has increased, and has made possible significant functions of management on the nodes in a distributed mode.

---

[4] These agents are not related with mobile agents which we will discuss later in this paper; to avoid confusion we will refer to this kind of agents with the term *snmp-agents*

Centralization is generally appropriate for the applications with a limited need for distributed control, do not require a frequent polling of MIB variables, and need only a limited amount of information. The typical example of this is the monitoring and the related displaying of some MIB variables. For example, the state of the interface of a router, or the state of a link involve only the query and the displaying of a limited number of MIB variables, and is therefore suitable for a centralized management.

On the other end we have some applications requiring a frequent polling of several MIB variables, which need to perform some computations on a large amount of information. An example could be the computation of a function that indicates the functionality level of the network, which must frequently detect the variations on a high number of MIB variables. In these cases, the monitoring and the control should be very close to the device. These drawbacks of the centralized approach are more evident in the periods of network congestion, when the manager's action is very important. In fact, during those periods, the management operations required for overcoming the congestion cause an increase in the traffic quite close to the congested part of network.

3 Distributed management

The disadvantages of a totally centralized architecture as the one described before have been admitted by the same organizations that had introduced it. A form of decentralization (provided in the first version of SNMP) is the mechanism of notification of asynchronous events. In fact, snmp-agents can send *traps* (that is, messages to the NMS) not as result of a request, but when some specific events occur. Since the basic purpose in the IETF was that of maintaining agents as simple as possible, the events that an agent can communicate to the NMS using traps are very simple: typical examples include the change in the state of a component (for example, from active to inactive, and vice versa). In any case, the task of snmp-agent is only the notification of the event: no management action can be performed locally, because any decision is made centrally by the NMS.

The SNMPv2 [11] introduces another decentralization feature: the concept of *proxy agent*, which leads to a structure of hierarchical management. A proxy agent is responsible for a set of devices, and the NMS sends the requests to it, when interacting with such devices. However, the SNMPv2 has not been used as widely as the first version, due to the few implementations available, and to the criticism made about the security model introduced.

Another approach to decentralization has been proposed by the IETF with the introduction of the RMON (*Remote MONitoring*) [12]. The RMON assumes the existence of appropriate devices called *probes*, whose task is to provide indications concerning the network traffic in a specific sub-network, and concerning the state of some devices. The RMON provides an approach oriented to the analysis of traffic, rather than to the device, as in the case of the ordinary snmp-agents. Anyway, the benefits of the RMON concern the *semantic compression* of data, preprocessing all the collected information, and sending only the significant ones to the NMS.

3.1 Approaches based on mobile code

The use of approaches based on mobile code for network management allows to overcome some limitations of current centralized management systems. In general, such approaches are based on the idea of reversing the logic according to which the data produced by network devices are periodically transferred to the central management station. Thanks to such technologies, the management applications can be *moved* to the network devices, thus performing (locally) some micromanagement operations, and reducing the workload for the management station and the overhead in the network [2].

This idea was expressed before developing the first systems based on mobile code. In fact, the idea of *management by delegation*, which is present in [6, 5], shows an architecture where the management station, sends some commands to the remote device (instead of limiting to the request for MIB variables), *delegating* the remote agent to the actual execution. Of course, remote devices include a so-called *elastic process runtime support*, which can provide for new services and dynamically extend the ones present in the device.

The approaches based on a mobile code may include the management by delegation as a particular case, since their field of application is wider, and they are more flexible.

In the following paragraphs we will examine the management functionalities that can take advantage from the adoption of a specific scheme based on the mobility of the code.

Use of Code on Demand The use of a paradigm based on *Code on Demand* considerably improves the flexibility provided by a network management system. In fact, the structure of the MIB in the SNMP approach (as well as the type of events that can be notified to the central station) is codified in the snmp-agent, and cannot be modified. The snmp-agent can in no way manage events defined by the user, nor it can act locally without involving the NMS. Maintaining management agents as simple as possible was one of the project restrictions followed while developing the SNMP, due to the limited computation capabilities of network devices. This was true until some years ago. Now, such devices have more powerful processors, and higher amounts of memory. However, we cannot think (this is not effective) of statically codifying all the management functions in an agent. In fact, a function (due to the specific nature of such applications) might be requested only on some devices. On other devices it could be seldom requested; on other ones it might be requested more frequently, while in other devices it could not be necessary. It is therefore clear that the static inclusion of such functions in an agent would lead to a considerable waste of resources.

The use of an approach based on Code on Demand increases the flexibility of the system, and maintains agents simple and *small* at the same time. Only when a function becomes necessary on a node, it can be *downloaded* from a *code server* and dynamically executed [9, 8]. In the same way, each software update requested by the agent can be obtained through this approach.

Use of Remote Evaluation The use of *Remote Evaluation* allows to deal with the issue of the bandwidth waste that occurs in a centralized system when micromanagement operations are needed.

In a traditional system, no matter which processing has to be performed on the data of the device, they have to be moved from the agent to the NMS, where they are processed. Conversely, thanks to the mechanism of Remote Evaluation, the actions to be performed can be developed and sent to the remote device, where they will be executed without generating traffic in the network (except the initial one for sending the code to be executed).

A typical example is the computation of the so-called *health functions* [5, 8, 4]. In general, by health function we mean an expression consisting of several elementary management variables. Each variable provides a particular measure concerning the device to be controlled. In traditional systems the computation of this function is performed by transferring the variables involved to the NMS (through polling operations at regular time intervals), and doing the computation on site. In this way, the higher the number of the variables involved (or the frequency at which such function has to be computed), the larger the amount of bandwidth occupied on the network (and the computational load produced on the NMS) will be. Even if this function can be codified within the agent, this is not convenient for the following reasons: the increase in the size of the agent and the inefficiency of the system due to the fact that a generic function might be seldom used.

By using a technology based on Remote Evaluation, we can compute these functions directly on the devices, and only when necessary. The code that performs such computations will be sent by the NMS and dynamically executed. This approach allows to obtain what is called *"semantic compression of data"*. In fact, generally the manager of a network is not interested in the single value of a MIB variable, but in aggregate values containing higher-level *"informations"*. We can therefore develop a system in which the manager writes its management functions (or they might be already available), and then invokes their remote execution on specific devices, when necessary.

Use of mobile agents The use of approaches based on mobile agents adds more benefits to the ones that can be obtained with Remote evaluation. In fact, in this case the action on a device is always expressly started by the NMS. Conversely, in the case of mobile agents, the ability to store the state during the movements allows to develop applications in which the agent moves from a device to another, performing the management functions required. Besides, the agent can be delegated the task of deciding where and when to migrate according to its current state, thus reducing the interaction with the NMS, and making processings more distributed.

The reduced interaction with the NMS is an advantage if we refer to situations in which the NMS has to manage different LANs and is linked to them by unreliable and/or low bandwidth links. In these cases, a traditional approach, which requires the exchange of several messages among the NMS and remote devices for a single operation, would mean delays in the execution of management operations, as well as their unreliability. Conversely, if we use mobile agents, once a given operation (which has been developed according to this new model as a mobile component) is sent to the remote LAN, it can be run there. Even if the agent's size increases, there will be no negative effect, since in a LAN we can always assume to have large bandwidths available.

4 Network management using MAP

In this section we introduce the agent system MAP [8] that we developed and implemented. MAP is a platform for the development and the management of mobile agents that gives all the primitives needed for their creation, execution, communication, migration, etc.

It has been entirely developed in Java, and this guarantees its total portability on the different hardware and software architectures. Java is also used as a language for programming agents: the programmer will not therefore need to learn several and/or specific languages, but only the API that are made available by the platform.

The migration facility implemented in MAP is a *weak migration*: the agents are able to carry the data state with them, but not the execution state. Besides, it integrates the paradigms of *remote execution* and *code on demand*, thanks to the exploitation of the Java mechanisms of classes dynamic loading. A graphic interface manages the agents in execution: this interface allows to verify which agents are active in the local server, and which ones are present in remote servers. Besides, through this interface we can trigger the execution of any agent present in the system, independently of where the agent code is resident.

In this section we show how, by using the services of the MAP platform, we have developed an agent-based application for performing the basic management operations of a network. The purpose of the implemented agents is to prove the effectiveness of an approach based on the mobility of the code for solving some of the issues described before. In particular, using the developed agents we are going to deal with the following issues:

- collection of information about the state of the network
- micromanagement of network devices through the monitoring of *health functions* defined by the user.

The application through which the actual management agents are triggered and run, has been structured as an agent (called *snmp-monitoring*). Through the graphical interface of this agent (shown in Figure 1) we can select the agents to be sent to the network, control the list of the active agents in the network, and evaluate the information collected by the agents that have finished their task.

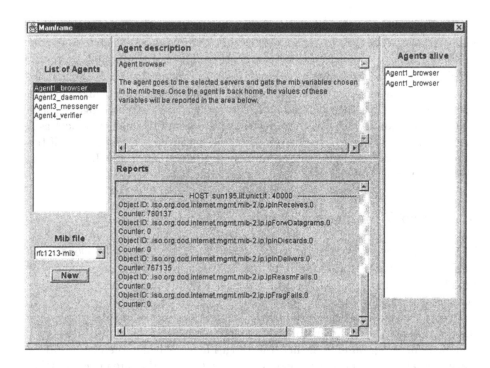

Fig. 1. Graphical interface of the application *snmp-monitoring*

According to the type of agent selected, an appropriate dialog box is viewed. Through this box, the typical parameters for a kind of agent can be specified. There are some advantages in structuring the *"snmp-monitoring"* application as an agent: the management application can be activated from any node where a MAP server is present, thus releasing the manager from the console of the NMS.

One of our purposes in the creation of the platform has been the complete integration and compatibility with the SNMP world. For this reason, in our system we have integrated and used some classes created by Advent [1], which implement the SNMP stack. In this way, we have had the opportunity to interact with the different nodes of the network through standard MIB variables. At the same time the developed framework allows us to monitor non-standard quantities (not defined by a MIB) defined by the user.

4.1 Agents for the management

We have developed the following basic types of mobile agents for the management:

- *browser* agent
- *daemon* agent
- *messenger* agent
- *verifier* agent

They perform very simple tasks, but the combination of their actions allows to perform even very complex actions of management. However, our purpose is not creating a complete management system, but showing how the ideas presented before can be easily implemented by our agents system.

Below we describe the details and the functionalities of each type of agent carried out.

Browser agent The *browser* agent collects some MIB variables from a set of nodes specified by the user. Both the variables of the mib-tree to be collected and the servers to be visited are selected through an appropriate dialog box (Figure 2) of the application *"snmp-monitoring"*. After being started, the agent reaches the first node to be visited, opens a Snmp local communication session, builds a pdu-request containing the MIB variables to be searched, waits for the reply from the Snmp daemon, and saves the data obtained in an internal structure. Then, if other nodes need to be visited, it reaches them, and repeats the procedure mentioned above. Otherwise, it returns to the platform from which it has been launched, where the results of the research are shown.

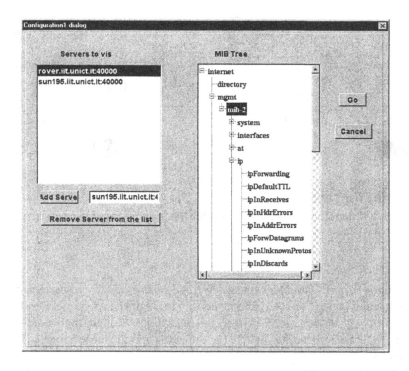

Fig. 2. Graphical interface of the browser agent

This agent realizes the functionalities of an extended MIB browser. In fact, in an ordinary MIB browser, after activating a connection with a specific node, we can view the contents of the MIB variables, by sending a Snmp request to the node in question for each variable. Conversely, through the browser agent, such requests are first given to the agent, which (by moving from a node to another) interacts with them locally and reports the global result to the initial station.

Daemon agent The *daemon* agent monitors a "health function" defined by the user. For starting the agent, the function to be computed and the node of the network (where the computation has to be done) must be provided. Then this agent moves to the node in question, where it records the value of the function: if the value is higher than a fixed threshold (defined by the user), a notification message is sent to the server from which the agent has departed.

Figure 3 shows the basic functioning mechanisms of the daemon agent. We can implement (by means of this agent) a mechanism of *generalized trap*, where the events in which the NMS is notified can be freely and very flexibly set by the user. Remote nodes (thanks to the use of the mobile code) do not need to have anything preinstalled.

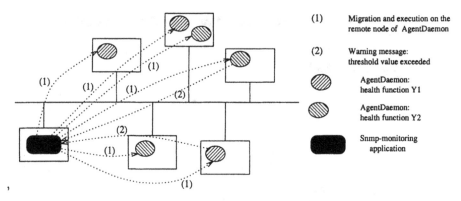

Fig. 3. Computing customized health functions by means of daemon agents

Messenger agent The two agents described before directly interact with the Snmp daemon present in the different nodes (through the Advent classes). Conversely, the *messenger* agent, during its migration through the nodes of the network, interacts with other agents for collecting specific information produced by them.

During the configuration we need to select the agents to be contacted and the servers where they have to be searched, and (if necessary) also the number of times the agent has to contact such agents. Thus, the messenger agent performs operations at a higher abstraction level than the mere retrieval of MIB variables.

In fact, since daemon agents can perform the computation of any function on the different nodes of the network, the messenger allows to collect such information, thus obtaining a general description about the state of the network.

The schematic behaviour of a messenger agent is graphically shown in Figure 4. It must be noticed that the model of the agents involved in this architecture is always the same. The functions to be computed on each node, which clearly depend on the specific environment where we operate, need to be changed for obtaining significant information about the network (below we will show the simple steps to be followed for inserting new monitoring functions in the system).

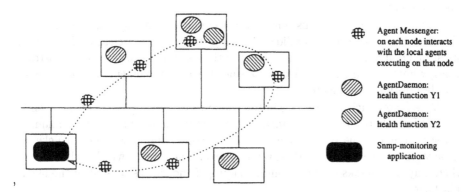

Fig. 4. Messenger agent: interaction with other system agents

Verifier agent The *verifier* agent does not perform an actual network management action. Its task is that of bringing important information to the base, which might be useful for further operations of network management. It visits the nodes selected during the configuration, and computes a function whose purpose is the evaluation of the presence of a specific property of the system visited (for example, we might think of the verification of a software version, or the available space on disk not below a fixed threshold, or the verification of some log files, etc.). The verifier agent then reports the server, from which it departed, the list of the nodes that correspond to this property.

4.2 Adding new functions to the system

In the agents described before we have referred to several functions that they can compute on the different nodes. The system proposed allows to develop and add new functions very easily. In fact, by respecting some very simple rules, any function can be inserted in the system. From the point of view of the implementation, an abstract class *Function* is present, which is a sort of template for building new functions.

This is the definition of this class:

```
abstract public class Function implements Serializable {

    abstract public void initfun (String mf);
    abstract public Vector calc();
}
```

A user who wishes to write his/her own monitoring function, only needs to extender this class, by implementing the two methods *initfun* (which is called only once before starting the function) and *calc* (which performs the actual computation). Besides, the user can access the MIB variables through appropriate classes in each node.

Then, in order to use this function in the management system, the name of the class must begin by *fun* followed by any string. Then a class of description must be created for each function. Even in this case we have created an abstract class, whose only method called *getDes* needs to be implemented. The name of the description class has to begin by *dfun*.

If we follow such simple rules in the configuration dialog boxes of the agents, we will be able to distinguish functions and descriptions from all the remaining classes present in the system. We will have the opportunity to store such classes in any node where the MAP server is active. The agents will be able to use them anyway, thanks to the dynamic loading of the classes present in the agent platform.

5   Conclusions

In this paper we have presented a system of network management based on mobile agents technologies. We have discussed the main benefits deriving from the adoption of these technologies, comparing and evaluating the applicability of the different programming paradigms, based on a mobile code.

By using the MAP agents platform, we have developed some agents for performing basic network management functionalities, including a generalized MIB browser. The possibility of customizing the network functions to be monitored is one of the most interesting aspects of the proposed approach. Future developments in this work concern the definition of more complex health functions, and of agents that could adapt their migration dynamically according to the type of task they have been requested.

Acknowledgments

This work has been carried out under the financial support of the Italian Ministero dell'Universitá e della Ricerca Scientifica e Tecnologica (MURST) in the framework of the MOSAICO (Design Methodologies and Tools of High Performance Systems for Distributed Applications) Project.

References

1. AdventNet. AdventNet SNMP tools. *http://www.adventnet.com/*, 1998.
2. M. Baldi, S. Gai, and G.P. Picco. Exploiting Code Mobility in Decentralized and Flexible Network Management. In *Proceedings of the First Int. Workshop on Mobile Agents (MA97)*, Berlin, Germany, April 1997.
3. J. D. Case et al. A Simple Network Management Protocol (SNMP). *RFC 1157*, 1990.
4. G. Goldszmidt. On Distributed System Management. *Proceedings of the IFIP International Symposium on Integrated Network Management*, 1993.
5. G. Goldszmidt and Y. Yemini. Distributed Management by Delegation. *Proc. of the 15th International Conference on Distributed Computing Systems*, 1995.
6. G. Goldszmidt, Y. Yemini, K. Meyer, M. Erlinger, J. Betser, and C. Sunshine. Decentralizing Control and Intelligence in Network Management. *Proc. of the 4th International Symposium on Integrated Network Management*, May 1995.
7. K. McCloghrie and M. Rose. Management Information Base for Network Management of TCP/IP-based Internets: MIB-II. *RFC 1213*, 1991.
8. A. Puliafito, O. Tomarchio, and L. Vita. MAP: Design and Implementation of a Mobile Agent Platform. *Journal of System Architecture*. to be published.
9. A. Puliafito, O. Tomarchio, and L. Vita. A Java-based Distributed Network Management Architecture. In *3rd Int. Conference on Computer Science and Informatics (CS I'97)*, Durham (USA), March 1997.
10. Marshall T. Rose. Network Management is Simple: you just need the Right Framework. *Proceedings of the IFIP Second International Symposium on Integrated Network Management*, 1991.
11. W. Stallings. *SNMP, SNMPv2, and CMIP. The practical guide to network-management standards*. Addison Wesley, 1993.
12. S. Waldbusser. Remote Network Monitoring Management Information Base. *RFC 1757*, 1995.

Market-Based Call Routing in Telecommunications Networks Using Adaptive Pricing and Real Bidding

M.A.Gibney[1], N.R.Jennings[1], N.J.Vriend[2] and J.M.Griffiths[1]

[1]Department of Electronic Engineering, [2] Department of Economics,
Queen Mary and Westfield College, University of London,
London E1 4NS, UK.
{M.A.Gibney, N.R.Jennings, N.Vriend, J.M Griffiths}@qmw.ac.uk

Abstract. We present a market-based approach to call routing in telecommunications networks. A system architecture is described that allows self-interested agents, representing various network resources, potentially owned by different real world enterprises, to co-ordinate their resource allocation decisions without assuming *a priori* co-operation. It is argued that such an architecture has the potential to provide a distributed, robust and efficient means of traffic management. In particular, our architecture uses an adaptive pricing and inventory setting strategy, based on real bidding, to reduce call blocking in a simulated telecommunications network.

1 Introduction

In telecommunications networks, call traffic is typically routed, through the network from source to destination, on the basis of information about the traffic on that path only. Therefore, path routing is carried out without regard to the wider impact of local choices. The main consequence of this myopic behaviour is that under heavy traffic conditions the network is utilised inefficiently: rejecting more traffic than would be necessary if the load were more evenly balanced. One means of performing such load balancing is to centrally compute optimal allocations of traffic across the network's paths using predictions of expected traffic [Bertsekas&Gallager87]. When such calculations have been completed, the network management function can configure the network's routing plan to make the best use of the available resources given the predicted traffic. However, as networks grow larger and involve more complex elements, the amount of operational data that must be monitored and processed (by the network management function) increases dramatically. Therefore in centralised architectures, management scalability is bounded by the rate at which this data can be processed [Goldszmidt&Yemini98]. In addition, there are a number of well known shortcomings with algorithms to compute optimal network flows; these include progressively poorer performance in heavily loaded networks and unpredictable oscillation between solutions [Kershenbaum93]. Furthermore, the very centralisation of the network management function provides a single point of failure; thus making the system inherently less robust.

For the above mentioned reasons, a decentralised approach to routing is highly desirable. In such cases, decisions based on more localised information are taken at multiple points in the system. The downside of this, however, is that the local decisions have non-local effects. Thus, decisions at one point in the system affect subsequent decisions elsewhere in the system. Ideally localised control would take place in the presence of complete information about the state of the entire system. Such a state of affairs would enable a localised controller to know the consequences of a choice for the rest of the network. However, there are two main reasons why this cannot be realised in practice. Firstly, the network is dynamic and there is a delay propagating information. This means that a model of the network state held at any one point is prone to error and difficult to keep up to date. Secondly, the scaling issues involved in making flow optimisation computations for the entire network (noted above) would obtain here also. Therefore, a system in which local decision making takes place in the presence of an incomplete view of the wider network is the only feasible solution for providing distributed control.

A promising approach that combines the notion of local decision making with concerns for the wider system context is that of agent technology [Jennings&Wooldridge98]. Agents address the scaling problem by computing solutions locally, based on limited information about isolated parts of the system, and then using this information in a social way. Such locality enables agents to respond rapidly to changes in the network state; while their sociality can potentially enable the wider impact of their actions to be coordinated to achieve some socially desired effect. Systems designed to exploit the social interactions of groups of agents are called multi-agent systems (MAS). In such systems, each individual agent is able to carry out its tasks through interaction with a small number of acquaintances. Thus, information about the extent of the system is distributed along with whatever functionality the MAS is designed to perform.

One agent-based technique that is becoming increasingly popular as a means of tackling distributed resource allocation tasks is market-based control [Clearwater96]. In such systems, the producers and consumers of the resources of a distributed system are modelled as the self-interested decision-makers described in standard microeconomic theory [Varian92]. The individual agents in such an economic model decide upon their demand and supply of resources, and on the basis of this the market is supposed to generate an equilibrium distribution of resources that maximizes social welfare. In market-based control, the metaphor of a market economy as a system computing the behaviour that solves a resource allocation problem is taken literally and distributed computation is implemented as a market price system. That is to say, the agents of the system interact by offering to buy or sell commodities at given prices [Wellman96]. In our case, such an approach has the advantage that ownership and accountability of resource utilisation are built into the design philosophy. Thus, market-based solutions can be applied to the management of multi-enterprise systems without forcing the sub-system owners to co-operate on matters affecting their own commercial interest.

Within this context, this paper describes a system to balance traffic flow through the paths of a logical network, based on the local action of agent controllers coupled with their social interaction as modelled by a computational market. The approach presented here builds upon the preliminary work reported in [Gibney&Jennings98] in that it shares the same architecture, roles and deployment model for the agents. However, to improve upon our earlier results we devised a new approach to the way that agents adapt their pricing and inventory strategies according to the outcome of individual market actions and the profitability of trading in the market. More generally speaking, this paper extends the state of the art in market-based control in the following ways. Firstly, it models a complex two-level economy, in which not only end users but also the internal components of the system compete with one another for resources. The rationale behind using a two-level economic model is to realise call admission control in the same framework as the network management function. This is novel, as market-based control has not previously been used to address two control issues in the same system. Particularly, having two kinds of market within the economy, with agents active in different roles in each of them, provides an elegant way to acquaint agents with one another. This architecture also provides an appropriate way to situate the intelligence of the system in a multi-enterprise network with self-interested enterprises. Secondly, a novel approach is adopted to pricing strategy. Our agents adapt to the outcomes of market interactions which use real bids and offers (i.e. agents state a price in an auction-like market and are then committed to buying or selling at that price in that session). This approach was adopted because it eliminates the lengthy series of interactions between agents that is required to calculate the equilibrium price in the market. Rather, we use real bidding and allow the agents to adapt their bidding behaviour to the outcomes of the auctions over time. Real bidding allows us to use more rapid (one shot) auction protocols as markets.

The remainder of the paper is structured in the following manner. We discuss the background and motivation for this work in Section 2. Section 3 describes the architecture of the system as a whole and the institutional forms of the possible interactions between agents. The design of individual agents is given in Section 4. Section 5 discusses the experiments carried out to evaluate the performance of the system. Finally, Section 6 details our conclusions on the work presented in this paper and discusses the open issues and future work.

2 Background and Motivation

Decentralized approaches to routing, usually in packet switched telecommunications networks, based on the interaction of controllers distributed through the network have existed for some time [Schwartz&Stern89]. However, agent based approaches extend this idea by modelling the interaction of distributed controllers as a social process. A number of agent based solutions have been proposed to the problem of load balancing in telecommunications networks. [Appleby&Steward94] make use of mobile agents roaming the network and updating routing tables to inhibit or activate routing behaviours. [Schoonderwoerd et.al.96] extend and improve this approach by using ant-like mobile agents that deposit "pheremones" on routing tables to promote efficient routes (the majority of ants use the efficient routes and the pheremones re-enforce this behaviour in other ants). [Hayzelden&Bigham98] employ a combination of reactive and planning agents in a heterogeneous architecture to reconfigure route topology and capacity assignments in ATM networks. All of these systems exhibit increased robustness and good scaling properties compared to centralised solutions. Indeed, in network environments with symmetric traffic requirements, ant-like agent solutions have even been shown to provide superior load balancing to both statically routed networks and a more conventional mobile agent approach [Schoonderwoerd et.al.96].

However, all the aforementioned approaches model networks as a single resource and therefore act to optimise utilisation of that resource. This makes sense because a poorly managed telecommunications network benefits no one. However, the main disadvantage of such a perspective arises when different telecommunications operators join their networks together (something which is an increasingly common trend). In such cases, if the different sub-network owners agree on a single, unified static network management policy, it is unlikely that this policy will benefit all their interests individually as well as collectively over time. We address this issue by modelling our agents as the resources and groups of resources that enterprises might own or lease in a multi enterprise environment.

Another increasingly common aspect of modern telecommunications deployment is the practice of enterprises in other sectors (banking and other traditional consumers of telecommunications services) leasing bandwidth from telecommunications providers to create virtual private networks in the short, medium and long term [Cisco99]. Again, this promotes the creation of multi-enterprise networks. In such environments each enterprise clearly has an incentive both to see that overuse does not degrade network performance and to make the greatest possible use of their network ownership. Since these parties cannot agree each traffic policy decision individually, conflicting incentives must be reconciled outside the traffic management domain. Typically this is achieved by allowing sub-network owners to set policy within the remit of their own resources [Stallings97]. However, the static nature of these policies and the conflict between them at sub-network interfaces often causes institutionalised under-use of the network as a whole.

Both of these trends suggest that telecommunications network management, once a centralised and monolithic undertaking, will increasingly benefit from an open, robust, scalable and inherently multi-enterprise approach. Therefore, one of the aims of this work is to use the multi-agent system paradigm to address the problem of multi-enterprise ownership of the network, while simultaneously addressing the problems of robustness and scalability. Against this background, the resource allocation problem in a network with multiple, non co-operating enterprises can be recast as the problem of reconciling competition between self-interested, information-bound agents. We conjecture that a market economy might be an effective mechanism for achieving this goal. Therefore we decided to implement our telecommunications network management framework using economic concepts and techniques.

3 System Architecture

The overall system architecture consists of three layers (Figure 1.). The lower layer is the underlying telecommunications infrastructure. The middle layer is the multi-agent system that carries out the network management function. The top layer is the system's interface to the call request software. More details of the rationale to this design are given in [Gibney&Jennings98]. The remainder of this section concentrates on the agent layer: describing the main components (section 3.1) and how they relate to one another (section 3.2).

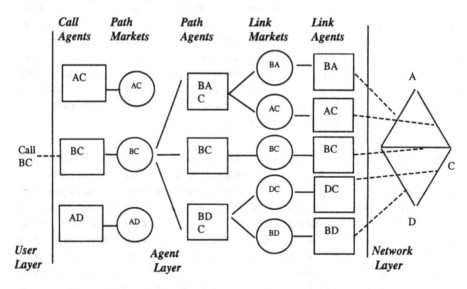

Fig. 1. System Architecture

3.1 The Agents and their Interactions

The system makes use of three agent types: (1) the *link agents* (section 4.1) that represent the economic interests of the underlying resources of the network, (2) the *path agents* (section 4.2) that represent the economic interests of paths across the network and (3) the *call agents* (section 4.3) that represent the interests of callers using the network to communicate. A *link agent* is used for every link in the network and is deployed at the entry node for that link. A *path agent* is used for each logical path in use across the network and is deployed at the source node for that path. Here we use three path agents for each source destination pair. Three is a reasonable number of alternate paths across which to share a single traffic requirement (alternate static routing systems commonly using three or fewer paths). A *call agent* is used for each source destination pair in the network and is deployed at the source of the traffic requirement that it represents.

The agents communicate by means of a simple set of signals that encapsulate offers, bids, commitments, and payments for resources. We couple the resources and payments with the offers and bids respectively. This reduces the number of steps involved in a transaction (committing agents to their bids and offers ahead of the market outcome), and so increases the speed of the system's decision making (an important consideration in this domain). To enforce these rules the interactions between the different agent types are mediated by means of market institutions (described in section 3.2).

An important notion in agent technology is that agents should be proactive (i.e. be able to anticipate the requirements of the environment and behave accordingly). In our system, we apply this concept to implementing a call routing mechanism that does not need to examine the network state before routing each call. Our path agents proactively determine how many calls they will be able to handle in advance, and seek to obtain the necessary resources to handle them. To be able to offer resources to callers pro-actively, the path agents lease bandwidth from the link agents over a period of time, paying installments on the lease at prices agreed on the link markets.

3.2 Market Institutions

Our system makes use of two types of market institution: At the *link market* (section 3.2.1), slices of bandwidth on single links (the fundamental resources of the system) are sold to path agents. At the *path market* (section 3.2.2), the slices of bandwidth on entire paths across the network are sold to call agents to connect calls.

3.2.1 Link Markets

Link markets are sealed bid double auctions. In this protocol the auctioneer receives two sets of prices in each trading period: bids for resources from buyers and offer prices from sellers, and computes the successful trades according to a set of rules defined for the auction. A sealed bid protocol was chosen because it provides a means to complete institutionally mediated bargaining in one shot. Therefore bargaining that would take an indeterminate time using iterated market institutions such as continuous double auctions can be completed almost instantaneously.

The resources exchanged at the *link markets* are the right to use slices of bandwidth on individual links, which when taken together, provide the necessary bandwidth to connect calls across paths. The link markets use a sealed bid double auction in which buyers and sellers periodically submit bids for individual units of the resource. In our case, buyers and sellers are constrained to their roles in the market by their position in the network. Thus, *path agents* need to buy resources from *link agents* to offer services to callers. The bids and offers are ordered from high to low, and low to high respectively. There will be a range of prices for which the market will clear at which the maximum amount of resources can be traded. Buyers bidding above these prices and sellers offering below it are allowed to trade. The buyers and sellers within this group are matched randomly so that the benefits of trade are assigned between successful agents randomly also. The trading price for each given transaction is determined at random in the range between the buyer's bid and the seller's offer. Notice that this procedure implies that no buyer will pay more than his bid, and no seller will get paid more than his offer. Moreover, the procedure implies that the total surplus realized in the market, a measure of the social welfare, is maximized because the benefits of trade are distributed randomly between all successful agents.

3.2.2 Path Markets

The path market is also a sealed bid auction. This is because it is a critical performance requirement of the system that the allocation of call traffic to paths occur almost instantaneously (so that callers are not kept waiting for calls to be established). This means the auction protocol has to be as short as possible. As before, the most efficient protocols in this respect are the single shot, sealed bid types. Since we have a single caller and multiple path agents offering resources, a single sealed bid auction is appropriate.

A buyer sending a service request message to the market initiates the auction. The auctioneer then broadcasts a request for offers to all agents able to provide the connection. All sellers simultaneously submit offers and the lowest one wins the contract to provide the connection. In this market, we experimented with two protocols: (i) The First Price protocol, in which the price at which the buyer and the seller trade is that of the highest bid submitted. (ii) The Vickrey, or second price auction protocol, in which the price at which the buyer and seller trade is that of the second highest bid submitted. We choose to experiment with two strategies because economic theory predicts that Vickrey auctions provide more competitive market outcomes, doing away with wasteful speculation by encouraging truth telling behaviour on the part of the participants [Varian95]. However, since we are using simple adaptive agents without speculative bidding strategies, we were unsure as to whether this factor would impact the overall behaviour of the system. To test the impact of this factor on the system as a whole, we implemented the market with both protocols and empirically tested the efficiency of each (section 5).

4 Designing Economically Rational Agents

The range of potential interactions is determined by the market institutions in which the agents participate. In both of our market types, agent communication is restricted to setting a price on a single unit of a known commodity. Therefore, agents set their prices solely on the basis of their implicit perception of supply and demand of that commodity at a given time. When a resource is scarce, buyers have to increase the prices they are willing to bid, just as sellers increase the price at which they are willing to offer the resource (mutatis mutandis when resources are plentiful). Here, agents perceive supply and demand in the market through the success or otherwise of bidding at particular prices.

4.1 Link Agents

A *link agent* has a set of n resources, slices of bandwidth capacity required to connect individual calls, that it can sell on the *link market*. At time t, the price to be asked for each of these units is stored in a vector $p_t = \{ p_t^1, ..., p_t^n \}$ with the range of possible prices being zero to infinity, $p_t^i \in [0, \infty >$ for each member of the vector $i = 1,..., n$ and each time period t. At time $t = 0$ the prices for each unit are randomly (uniformly) distributed on $[0, H]$ where H is the initial upper limit on prices asked. When x units have been allocated, the remaining $n - x$ units are offered to the link market for sale simultaneously. Suppose that of the $n - x$ units offered for sale in a given period t, the m units with the lowest prices are successfully sold. The prices in the vector are updated as follows:

$$p_{t+1}^i = p_{t}^i, \text{ for } I = 1,..., x \tag{1}$$

$$p_{t+1}^i = p_{t}^i \times (1 + \varepsilon) \text{ for } i = x+ 1,...,m \tag{2}$$

$$p_{t+1}^i = p_{t}^i \times (1 - \varepsilon), \text{ for } i = x+m + 1,...,n \tag{3}$$

$$\text{where } \varepsilon = U(0, \sigma) \tag{4}$$

Thus the link agent increases or decreases the price of any unit by a small amount ε after each auction. Here ε is obtained from a uniform random distribution between zero and σ (here 0.1). If previously allocated units are released by the path agent, they join the pool of unallocated units and the price vector is re-ordered to reflect this. This approach was chosen so that the prices of each resource on the link, taken together, should adapt to the demand on the network to carry traffic.

4.2 Path Agents

A *path agent* acts as both a buyer of link resources and a seller of path resources. Their buying behaviour is detailed in section 4.2.1, and their selling behaviour in section 4.2.2. In general, path agents wish to buy resources cheaply from link agents and sell them at a profit to end consumers. To do this, they bid competitively to acquire resources that they then sell on to callers, at a price not less than that paid for them. The path agent tries to maximize its profits by adjusting its inventory and sales behaviour on the basis of the feedback it receives from the market. The mechanism by which the path agent decides what resource level to maintain is described in section 4.2.3.

4.2.1 Buying Behaviour

A *path agent* actively tries to acquire resources (units of link bandwidth needed to connect a call), across the chain of links that it represents. It does this by placing bids at each of the link markets at which the resources it needs are sold. The agent retains a vector of prices that it is willing to pay for resources on each of the links that constitute its path. The agent's strategy is to try to equalise its holding of resources across each of those links; uneven resource holdings have to be paid for but cannot be sold-on or bring in any revenue because they do not constitute complete paths. The *path agent* tries to maintain its resources at a level w that is discovered through hill climbing adaptation [Russell&Norvig95] to the behaviour of the market (section 4.2.3). The most profitable value of w is obtained by adjusting it according to changing profit during ongoing buying and selling episodes. When x units have been acquired, the path agent bids for the remaining w -x units on the link market simultaneously. Suppose that, for a given link, of the $w - x$ units that the path agent bids for at time t, the m units with the highest prices are successfully acquired. The prices in the vector are updated as follows (using ε as defined previously in section 4.1).

$$p^i_{t+1} = p^i_t, \text{ for } i = 1, ..., x \tag{5}$$

$$p^i_{t+1} = p^i_t \times (1 - \varepsilon) \text{ for } i = x+1, ..., m \tag{6}$$

$$p^i_{t+1} = p^i_t \times (1 + \varepsilon), \text{ for } i = x+m+1, ..., w \tag{7}$$

This price setting mechanism was chosen because it allows the path agents to adaptively determine prices for individual link resources. The price bid for each resource should be as low as possible without failing to win the resource in the auction. Therefore the agent makes a bid for each resource that it needs separately. If a bid fails, the agent increases the price it will bid at the next auction (in order to increase its likelihood of winning the resource). If a bid succeeds, the agent reduces the price it bids for that resource in subsequent auctions (in order to avoid paying over the market price).

4.2.2 Selling Behaviour

A *path agent* will offer to sell a path resource whenever an auction is announced by the path market and it has an appropriate path resource to sell. The price asked is determined by the cost of acquiring the underlying link resources and the outcomes of previous attempts to sell. Let p_t be the price of a path resource at time t (the time of the auction) ranging from the cost to acquire the resource to infinity, $p_t = [cost, \infty >$. The first time the agent offers a resource for sale, it offers it at a price given by $p_t = cost \times (1 + \varepsilon)$ in order to sell at a profit (using ε as defined previously in section 4.1). Subsequently the offer price is given by:

$$p_{t+1} = Max(\ p_t \times (1 + \varepsilon\ ),\ cost) \text{ if last offer was successful} \tag{8}$$

$$p_{t+1} = Max(p_t \times (1 - \varepsilon\ ),\ cost) \text{ if last offer was not successful} \tag{9}$$

This price function ensures that the path agent never sells a resource for less than it paid to acquire it in the first place. Given its inventory level (see section 4.2.3), the agent attempts to maximise its income. The price bid for each resource should be as low as possible without failing to win the resource in the auction. Therefore, the agent increases the price it asks whenever it is successful, and decreases it whenever it is unsuccessful. This means the agent adapts its price to the level of competition, as it perceives it from the outcomes of previous auctions.

4.2.3 Inventory Level

The resource levels of the various path agents determine the maximum flows available for traffic on individual paths. In our case, the system design philosophy is to have the individual paths determine their own optimal resource levels. When this is achieved, balancing the load in the network as a whole becomes an emergent property of the social interactions of the agents. Path agents act in response to the economic pressures exerted on them by their consumers, competitors and suppliers. Therefore, we choose to have path agents discover their own optimal flows by adaptation to economic conditions as they perceive them (through interactions in the markets in which they compete). Path agents are both *buyers* and *sellers* that attempt to maximise their profit through trade, where profit is the difference between their revenue (from selling path resources to callers) and their expenditure (cost of acquiring and holding onto resources). In order to maximise profit agents must have an inventory level that is optimal for them, in the competitive environment in which they find themselves. Therefore our path agents adapt their inventory levels to the profits they earn through their interactions with the market. This is implemented by having the inventory level of individual path agents climb the hill of their profits.

In more detail, let R_t be the profit of an agent at time t and I_t be the desired resource inventory of that agent. If profit has increased since the last market interaction ($R_t > R_{t-1}$) and the last change corresponded to an increase in desired inventory level ($\Delta I_t > 0$), the new desired inventory level is increased by one resource unit. If the last change in desired inventory was negative ($\Delta I_t < 0$) then desired inventory is reduced by one unit. However, if profits have fallen, ($R_t < R_{t-1}$) and the last change was positive ($\Delta I_t > 0$) we decrease the desired inventory level; If the last change was negative ($\Delta I_t < 0$) then the desired inventory level will be increased. When decreasing the desired inventory level, the agent chooses to give up the most expensive of its link resources (that are not allocated to a call). This strategy reduces the agents inventory rental cost by the largest amount possible in a single time step.

4.3 Call Agents

Call agents initiate the auctions at which path agents compete by signalling that they wish to buy resources to a given source destination pair. In doing so they set a maximum price that they will not exceed to make a call. This puts downward pressure on the offers made by path agents to provide resources across whole paths, thereby anchoring the system.

5 Experimental Evaluation

Our experiments were designed to test three hypotheses. Firstly, we wanted to know whether market-based systems can compete with static routing algorithms in terms of call routing performance (section 5.1). Secondly, we wanted to know if our system uses the network efficiently (i.e. does it use the best routes) whenever possible, allowing for congestion (section 5.2). Thirdly, we wanted to test if the system discriminates in its choice of routes between paths that would be indistinguishable from one another to a conventional static routing algorithm, without expected traffic predictions (i.e. they differ only in their proneness to congestion) (section 5.3).

5.1 Performance Evaluation

In terms of performance, we sought to address two fundamental questions. Firstly, can our market-based control system perform as well or better than a conventional system? Secondly, what is the effect of using a Vickrey auction protocol rather than a first price auction protocol at the path market?

In a series of experiments we tested the efficiency of our market-based control mechanism (using both first price and Vickrey auctions as path markets) against a static routing mechanism. The static routing mechanism used a number of paths (three) between each source destination pair and routed traffic to these in order of path length, using longer paths when the shorter ones became congested. Here efficiency was measured as the proportion of calls successfully routed through the network as a percentage of the total number of calls offered. The experiment was configured to simulate a small irregularly meshed network of 8 nodes with link capacities sufficient for 200 channels. Calls arrived on average approximately every 5 seconds, were routed between a randomly chosen source destination pair, and lasted an average of 200 seconds. Call arrival and call duration were determined by a negative exponential time distribution function, with U (0, 1) being a uniform random distribution between 0 and 1. The inter-arrival time between calls and the call duration were calculated using the formula: $f(x) = -\beta \ln U$, where β was the mean inter-arrival time and mean call duration respectively. The simulation was allowed to run for 20,000 seconds in each case. The traffic model was chosen to simulate a realistic call arrival rate and duration. The network dimensions where chosen to reflect a small network under heavy load.

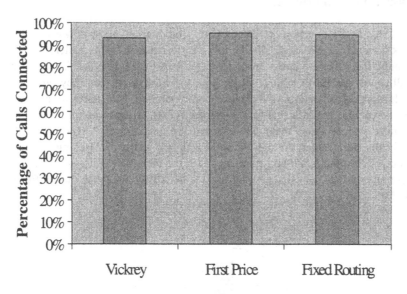

Fig. 2. Performance of Market-Based and Static Routing

These results show that similar levels of performance are obtained using the market-based control mechanism (95.4% of calls connected) and static routing (94.8% of calls connected) (Figure 2.). It is interesting to note that, contrary to our original hypothesis, using a Vickrey auction for the path market did not improve upon the results obtained using a first-price auction. One possible reason for this is the way in which the path agents in our system adapt their pricing strategy to market outcomes. Vickrey auctions are designed to make the markets more efficient by making counter speculation between competing agents wasteful. However, with simple adaptive agents being used here such speculation does not occur, so the effect should not be significant.

The ability of our system to perform as well as a static routing mechanism should be taken as a positive result. As well as matching the performance of conventional routing techniques, our market-based approach has a number of distinct advantages for network operators and users. Firstly, it provides an architecture that is open to deployment in multi-enterprise environments without the inefficiency of static internetworking policies at sub-network interfaces. Secondly, our system is scalable in that no agent has to know the address of a significant number of peer agents or possess a map of the entire network. Thirdly, our system allows a much quicker response to call requests because the call routing process does not need to obtain information from the wider network at call set up time. In our case a call request can be processed and the call dispatched to a path (or refused) solely on the basis of information present at its source. This is achieved by having path agents that pro-actively determine their capacity to handle traffic, rather than waiting for call requests before processing.

5.2 Resource Utilisation Efficiency

In addition to the raw connection performance, it is important to know how effectively the network's resources are used. This is important because both the callers and the system benefit from routing calls through the shortest path when one is available. Shorter call routes use fewer system resources than longer ones, and they provide a service with less delay to end users. To assess our system's performance, we analysed the relative percentages of calls that were assigned to first, second and third best paths (by the number of links which make up the path). The data (Figure 3.) shows that the majority of calls were routed through the most efficient routes: 61% using the most efficient route, 25% the second best route and 15% the third best. Thus not only does our system route most of the calls it is presented with, it also makes the most efficient routing choices.

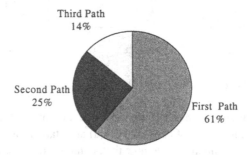

Fig. 3. Utilisation of First, Second and Third Shortest Paths by Market-Based Control

5.3 Congestion Discrimination

One of the claims we made for market-based approaches, is that good system level choices can emerge from local choices, that are influenced by information about the social context   (obtained through interaction). To explore this hypothesis, we examined the performance of our system in cases that are indistinguishable from a local perspective. Thus, we focussed on source-destination pairs where all the routes are of equal length. In such cases, an alternate routing mechanism cannot decide between these paths without some notion of congestion through the whole network, which cannot be calculated and propagated in real time (recall the discussion of section 2). Alternate routing mechanisms can assign traffic to paths probabilistically, so that statistically, over time more traffic is routed to less congestion prone paths . However, this method is dependent on the accuracy of past measured traffic as a predictor of future traffic patterns. With our approach, we believe such discrimination would emerge from the competitive nature of the market place. The reason for this hypothesis is that while path agents for paths of equal route length have to obtain the same number of resource slices, some have to obtain the more congestion prone of those slices. By definition, the more congestion prone resource slices are traded in the more competitive (and hence more highly priced markets). All other things being equal, the profitability of selling these paths will be lower because of the higher costs. With lower profitability comes a lower inventory level and fewer calls being routed via that path.

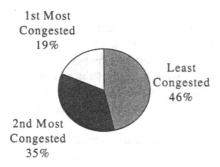

Fig. 4. Utilisation of paths in order of congestion by market-based control

We wanted to test whether the market-based control mechanism is able to discriminate between paths on the basis of congestion cost in real time. In order to investigate this, we examined the source-destination pairs in our network configuration (nine of them) for which all (three) known paths consisted in an equal number of links. We plotted the percentage of calls routed to each of the three paths in (reverse) order of congestion and took the average of these values across all nine source destination pairs (Figure 4.). Our results clearly show that the market mechanism is able to distinguish between congestion costs entailed in routing across paths of otherwise equal length and assign calls to the least congested path most of the time.

6 Conclusions

We have described the design and implementation of a market-based system for call routing in telecommunications networks. Our system performs comparably with a static routing approach in terms of the percentage of the calls that are connected. However, from an architectural point of view, the market-based approach represents an improvement on static, centralised systems for a number of reasons. Firstly, it provides a platform for implementing network traffic management in a multi-enterprise internetwork. Secondly, it does not rely on a centralised controller to compute network reconfigurations, making the network management function robust to failure. This means that the system described in this work is architecturally more robust than an equivalent network with a centralised control mechanism. It is important to distinguish between this sense of robustness, and robustness as an empirical measure of the performance of the system in the event of the failure of a component of the system (or the agent process managing it). The performance of a decentralised control system degrades with the loss of controlling processes, while a centrally controlled one would be without control if the controlling process were to be lost. The study of the degradation of control efficiency in the system described here has been left to future work. Thirdly, no agent needs to know of the existence of more agents than there are links in the paths of the network (making the agent acquaintance databases compact and the whole system more scalable). Fourthly, there is no requirement to test the network state at call set up time, making the call set up procedure faster and more robust. Fifthly, the cost of each call to the network and the proportion of that revenue owing to each of the enterprises involved in carrying that call can easily be computed from information available to the user terminal equipment when the call is made, thus making call charging more efficient.

The results presented in this paper show that our market-based system performs the call routing and network management tasks adequately. However, the function used to determine the inventory level of path agents is quite simple and responds reactively to burstiness in the call arrival rate; this may be inducing unwanted oscillation in the path inventory parameter which may be adversely affecting performance. We intend to experiment with this function, and the parameters that govern its behaviour, to determine the impact of our choices on the performance of the overall system, and to see if that performance can be improved upon.

Acknowledgement

This work was carried out under EPSRC grant No.GR/ L04801.

References

[Appleby&Steward94] S. Appleby & S. Steward, Mobile Software Agents for Control in Telecommunications Networks, BT Journal of Technology 12 - 2. (1994) pp 104 - 113

[Bertsekas&Gallager87] D. Bertsekas & M. Gallager, Data Networks, Prentice Hall International , Inc. (1987)

[Cisco99] Cisco Systems Inc., A Primer for Implementing a Cisco Virtual Private Network (VPN), Cisco Systems Inc. (1999)

[Clearwater96] S. H. Clearwater, Market-Based Control A Paradigm for Distributed Resource Allocation, Ed. S. H. Clearwater, World Scientific Press. (1996)

[Gibney&Jennings98] M. A. Gibney & N. R. Jennings, Dynamic Resource Allocation by Market-Based Routing in Telecommunications Networks, in Proceedings of Intelligent Agents for Telecommunications Applications 1998, Springer Verlag. (1998) pp 102 - 117

[Goldszmidt&Yemeni98] G. Goldszmidt & Y. Yemini, Delegated Agents for Network Management, in IEEE Communications Magazine, March 1998, IEEE (1998) pp 66 -70

[Hayzelden&Bigham98] A. Hayzelden & J. Bigham, A Heterogeneous Multi-Agent Architecture for ATM Virtual Path Network Resource Configuration, in Proceedings of Intelligent Agents for Telecommunications Applications 1998, Springer Verlag. (1998) pp 45 -59

[Jennings&Wooldridge98] N. R. Jennings & M. R. Wooldridge, Agent Technology Foundations, Applications and Markets, Springer Verlag. (1998)

[Kershenbaum93] A. Kershenbaum, Telecommunications Network Design Algorithms, McGraw-Hill International Editions. (1993)

[Russell&Norvig95] S. J. Russell & P. Norvig , Artificial Intelligence: A Modern Approach, Prentice Hall, Inc. (1995)

[Schoonderwoerd et.al.96] R. Schoonderwoerd, O. E. Holland, & J. Bruten, Ant-like agents for load balancing in telecommunications networks, in The First International Conference on Autonomous Agents, ACM Press. (1997)

[Stallings97] W. Stallings, Data and Computer Communications, Prentice Hall International , Inc. (1997)

[Schwartz&Stern89] M. Schwartz & T. E. Stern, Routing Protocols, in Computer Network Architectures and Protocols (Second Ed.), Ed. C.A. Sunshine, Plenum Press, New York / London (1989)

[Varian92] H.R. Varian, Microeconomic Analysis (Third Ed.), W.W. Norton & Company Inc. (1992)

[Varian95] H.R. Varian, Mechanism Design for Computerised Agents, Proccedings of the 1995 Usenix Workshop on Electronic Commerce. (1995)

[Wellman96] M. Wellman, Market Oriented Programming: Some Early Lessons, in Market-Based Control a Paradigm for Distributed Resource Allocation, Ed. S. H. Clearwater, World Scientific Press. (1996) pp 74 -95

Co-operating Market/Ant Based Multi-agent Systems for Intelligent Network Load Control

Brendan Jennings[a] and Åke Arvidsson[b]

[a] Teltec Ireland, Dublin City University, Dublin 9, Ireland
Brendan.Jennings@teltec.dcu.ie
[b] Ericsson Core Network Products, Soft Center, S-372 25 Ronneby, Sweden[1]
Åke.Arvidsson@uab.ericsson.se

Abstract. Recent years have seen increases in the number, complexity and usage of telecommunications services, many of which are realised by systems based on the Intelligent Network (IN) architecture. As the volume of traffic carried by INs has increased there has been a realisation that flexible and efficient load control strategies are required to ensure that Quality-of-Service levels meet desired targets. In this paper we present an agent-based IN load control strategy, realised by two co-operating multi-agent systems making use of Market-based Control and Ant Colony Optimisation paradigms respectively.

1 Introduction

Rapid technological advances have encouraged ever-greater usage of tele-communications services, both in terms of the number of users and the information volumes that they produce. Continued growth in the market is welcome from the point of view of network operators and service providers, however the task of meeting customers' availability and service quality demands is becoming increasingly challenging. Convergence of fixed, mobile and Internet networks, together with the expected introduction of service delivery and management platforms based on distributed object technology is resulting in an increasingly complex, interconnected infrastructure. This infrastructure will require network management systems that are more responsive, adaptive, proactive, and less centralised then those currently deployed. Since these are properties of *agents* and *multi-agent systems* many in the telecommunications research community have recognised that agent-based technology has the potential to offer a timely solution to the growing problem of designing efficient and flexible management systems.

In this paper we discuss the application of agent technology to load control in Intelligent Networks (INs), focussing in particular on how separate multi-agent systems may co-operate to fulfil different goals within the context of an overall control strategy. The paper is structured as follows: §2 provides a brief introduction to IN load control and discusses the potential advantages of agent-based solutions; §3

---

[1] The work was carried out on behalf of The Department of Software Engineering and Computer Science, The University of Karlskrona/Ronneby, S-372 75 Ronneby, Sweden.

presents a generalised IN scenario used as the basis for development and evaluation of the multi-agent systems and outlines goals for their combined operation; §4 and §5 specify two multi-agents systems, based on Market-based Control [1] and Ant Colony Optimisation [2] paradigms respectively; §6 shows how the two multi-agent systems can co-operate to provide an effective load control; finally §7 draws some conclusions and outlines future work items.

2 Intelligent Network Load Control

The Intelligent Network architecture was developed as a means to introduce, control and manage services rapidly, cost effectively and in a manner not dependent on equipment/software from particular equipment manufacturers. The In standards specify four network element types: Service Switching Points (SSPs), Service Control Points (SCPs), Service Data Points (SDPs) and Intelligent Peripherals (IPs). These elements typically communicate with each other via a Signalling System No.7 (SS.7) network. SSPs facilitate end user access to services by means of trigger points for detection of service access codes. SCPs form the core of the architecture, they receive service requests from SSPs and execute the relevant service logic. SCPs are assisted by SDPs, which store service/customer related data, and by IPs, which provide services for interaction with end-users (for example automated announcements). A much more complete description of the IN architecture can be found in [3].

Traditional approaches to IN load control are 'node-based' in nature – they are centred on protection of the processors of individual SCPs; a discussion and comparison of a number of strategies of varying sophistication can be found in [4]. We have argued elsewhere [5] that, given current and future increases in number and usage of IN services, node-based strategies alone cannot guarantee that desired performance levels are consistently achieved and that therefore more 'network-based' solutions are required. Network-based load control strategies differ from traditional approaches in that they seek to optimise the usage of the resources of the network as a whole, rather then on a localised basis. Of necessity this requires manipulation of a large amount of information from throughout the network, thus strategies of this type appear suited to the types of distributed solution normally associated with multi-agent systems.

We note that flexible and adaptable network load control strategies could certainly be implemented using standard software engineering techniques, however we believe that there are many advantages to be gained through the adoption of an agent-based approach:

- *Methodology:* the agent paradigm encourages a knowledge/information centred approach to application development, thus it may provide a useful methodology for the development of control strategies that require manipulation of large amounts of data collected throughout a network;
- *Agent Communication Languages:* much work in the agent research community has focussed on development of advanced communications languages that, for example, allow negotiation in advance on the semantics of future communications. This degree of flexibility is not present in traditional communications protocols and would be of use in realising strategies that adapt to dynamic network

environments where, for example, traffic patterns change due to introduction/withdrawal of services;

- *Adaptivity:* agents may have adaptive behaviour allowing them to learn about the normal state of the network and better judge their choice of future actions. A substantial body of work relating to learning behaviour in the context of agent systems exists and could be leveraged to incorporate learning behaviour into a load control strategy;
- *Openness:* the agent approach is very open, in that individual agents may exchange data and apply it in different ways to achieve a common goal. This means that equipment manufacturers could develop load control agents tailored to their own equipment, but which could still effectively communicate with agents residing in other equipment types;
- *Flexibility:* the approach may facilitate a 'plug-and-play' architecture whereby an agent associated with a recently introduced piece of equipment can easily incorporate itself into the agent community and learn from other agents the range of parameters that it should use for its load control algorithms.

In summary it can be said that investigation of the use of agent technology for network load control is an attractive research topic as it may provide a useful knowledge-centred methodology for building flexible, adaptable, scalable, open and robust solutions for what is an inherently distributed problem.

3 Network Scenario and Agent System Goals

Increasing demand and the entry of new service providers into the market is necessitating the introduction of a greater number of SCPs into existing INs. This is leading to the situation where an SSP may be able to choose between a number of different SCPs that support the desired service(s). The agent-based load control strategies addressed in this paper are targeted towards this type of IN environment. To develop the strategies the generalised network scenario illustrated by Fig. 1 below has been used.

As Fig. 1 shows that network scenario contains I SCPs and K SSPs, where I is assumed to be less than K. It is assumed that the SCPs all support the same set of J services classes and all user service requests can be directed by the receiving SSP to any of the SCPs. SCPs incorporate locally the databases needed for completion of all service logic processing, thus SDPs not need to be considered, similarly SSPs are assumed to incorporate locally the functionality of IPs. IN physical elements communicate using an SS.7 network, which utilises a mesh shaped configuration of S Signalling Transfer Points (STPs). It is assumed that no messages are dropped whilst traversing the SS.7 network. Additionally we assume that the SCPs are the bottleneck in the system – no overloads occur in either the SSPs or the SS.7 network. The differences in the number of STPs that must be traversed by messages travelling between different SSP/SCP pairs allows the modelling physical proximity in the network – a message travelling from a given SSP will have to traverse more STPs for a 'far away' SCP then a 'nearer' SCP.

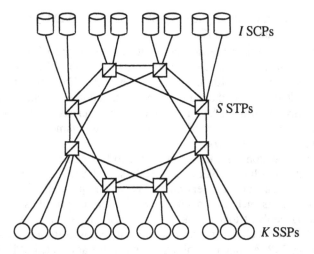

Fig. 1. Generalised Intelligent Network Scenario

The multi-agent systems (MASs) that will be described in the remaining sections have been designed to control the traffic load in the type of environment described above in an effective and efficient manner. More specifically they have been designed with the following goals in mind:

Goal 1. SCP load levels remain as close to a specified target load without ever exceeding it;

Goal 2. Network profit should be maximised;

Goal 3. Response delays for network users should be minimised;

Goal 4. The mechanism should be fair in its treatment both of services and users – accepted service sessions should experience similar response delays, regardless of service type or location of the user in the network;

Goal 5. Load should be balanced equally over all SCPs in the network;

Goal 6. Load on the SS.7 network should be minimised (through having SSPs select 'near' SCPs for its service requests where possible) and balanced over all STPs/links.

It is to be noted that some of these goals may be somewhat conflicting – for example having SSPs always select near SCPs (*Goal 6*) may mean that distant SCPs are under-utilised, thus load may not be balanced between all SCPs (*Goal 5*). The following sections will describe Market/Ant based MASs and indicate which of the above goals each would be able to meet.

4 Multi-Agent System for 'Co-operative Market' Control

In this section we will describe a 'Co-operative Market' multi-agent system designed to control allocation of SCP processing capacity in the network scenario described above. The agents utilise 'Market-Based Control,' a paradigm for distributed resource allocation based on application of concepts from economic science. Market-Based

Control has been successfully applied to problems such as ATM bandwidth allocation [6] and power load management [7].

The agents in the system realise load control through use of tokens – these are 'sold' by *Quantifier* agents associated with 'providers' (the SCPs) and 'bought' by *Allocator* agents associated with 'customers' (the SSPs). The amount of tokens sold on behalf of an SCP controls the load offered to it, and the amount of tokens bought on behalf of an SSP determines how many IN service requests it can accept. 'Trading' of tokens (in an 'auction') is carried out such that the common good is maximised, hence the term *co-operative market*.

To describe the auction algorithm in detail the following notation is introduced. Let i, j, and k denote an arbitrary SCP, service class, and SSP respectively and let $r(j)$ denote the profit generated by servicing a class j request. Profit values may be associated with factors such as generated revenue, service priorities or goodwill. All SSPs maintain IJ pools of tokens, one for each SCP and service class pairing. Each time SCP i is fed with a class j request one token is removed from the associated pool at the originating SSP. An empty pool indicates that the associated SSP cannot accept more requests of the associated class. Pools are refilled at auctions, which take place every T_{auct} time units (typically, T_{auct} would be on the order of tens of seconds). In the most basic version of the multi-agent system, which we will describe first, the auction algorithm is executed centrally by a *Distributor* agent using 'bids' (received in the form of messages every T_{auct} time units) from all the Allocators and Quantifiers in the system.

Quantifier bids consist of the unclaimed processing capability over the coming period of T_{auct} time units and the processing requirements for each service class respectively. Let c_i, $p_i(1)$, ... ,$p_i(J)$ denote the bids from Quantifier i. These bids can be given in any appropriate unit, for example the supply c_i may be the number of microseconds the associated SCP is willing to spend on new service requests over the coming T_{auct} time units, while the prices $p_i(j)$ would then be the number of microseconds associated with processing a request for a class j service. Allocator bids consist of the number of IN service requests expected over the coming period of T_{auct} time units for each service class respectively. Bids from Allocator k are denoted by $q_k(1)$, ... ,$q_k(J)$.

The auction algorithm maximises the expected overall utility, measured as total profit, over the next T_{auct} time units by maximising the expected marginal utility, measured as gain over cost, for every token issued. During the auction, a record is kept of the amount of capacity (in terms of tokens issued) sold on behalf of each Quantifier – let m_i be the processing capacity sold on behalf of SCP i. Similarly, a record is kept of the satisfied demands (in terms of tokens granted) of each Allocator – let $n_k(j)$ denote the number of tokens granted to SSP k for class j services.

The expected marginal gain $u_k(j)$ associated with allocating an additional class j token to Allocator k is defined as the profit associated with consuming it times the probability that it will be consumed over the auction interval. The latter may be obtained by interpreting the expected number of service requests $q_k(j)$ as the average of a Poisson distribution and we obtain:

$$u_k(j) = r(j) \sum_{w=n_k(j)+1}^{\infty} q_k(j)^w / w! e^{-q_k(j)} \tag{1}$$

The expected marginal cost $v_i(j)$ associated with Quantifier i issuing an additional class j token is expressed as the utility it represents to the Allocators. We get a measure of this by summing over all services and Allocators:

$$v_i(j) = \sum_{j'=1}^{J}\sum_{k'=1}^{K} \frac{p_i(j)}{p_i(j')} u_{k'}(j') \qquad (2)$$

Let an (i,j,k)-allocation refer to assigning a token from SCP i with respect to service class j to SSP k. The expected marginal utility $w_{i,k}(j)$ of such an action is the gain to cost ratio:

$$w_{i,k}(j) = u_k(j)/v_i(j) \qquad (3)$$

We note that a high marginal utility indicates that a large gain is obtained at a small cost. The triple i, j, and k which maximises the marginal utility therefore indicates the allocation which generates a maximum return on the spending.

Fig. 2 below illustrates the operation of the auction algorithm for the simple case where there is only one service class ($J = 1$) supported by the network. In step (1) Quantifiers and Allocators submit their bids to the Distributor, which then runs the auction process – step (2). In the diagram dark circles represent tokens, whereas light circles represent the expected number of required tokens; the auction algorithm assigns tokens to token requests in the stepwise fashion described below. Once the auction completes the values of token assignments are reported to the Allocators – step (3), which use them to admit service requests in the next time period.

The steps in the algorithm are as follows: the auction is initiated by setting $m_i = c_i$ for all i, $n_k(j) = 0$ for all j and k after which initial marginal gains $u_k(j)$ and costs $v_i(j)$ are computed for all for all i, j, and k. The steps in the algorithm are as follows: the auction is initiated by setting $m_i = c_i$ for all i, $n_k(j) = 0$ for all j and k after which initial marginal gains $u_k(j)$ and costs $v_i(j)$ are computed for all for all i, j, and k. During

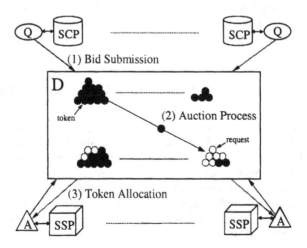

Fig. 2. Co-operative Market Load Control

the actual auctions tokens are handed out one by one from SCP i for service class j to SSP k, where i, j, and k are chosen to maximise the current values of $w_{i,k}(j)$. After each handout the remaining resources m_i, the satisfied demands $n_k(j)$, and the marginal gains $u_k(j)$ and costs $v_i(j)$ are updated. Resources m_i are updated by subtracting the value of the issued token $p_i(j)$, demands $n_k(j)$ by adding one token, and gains and costs as shown above. The auction is completed when no more resources remain, i.e. when $m_i = 0$ for all i. At this point the Distributor returns the results to the Allocators – each Allocator receives an IJ-dimensional matrix of token allocations. The order in which the tokens are used is at the discretion of the Allocator. The simplest mechanism would be to take each SCP in sequence, start to use all the tokens allocated for that SCP and when they are exhausted used the tokens for the next SCP; however, as will be discussed in §6 more desirable approaches are possible. It should be noted that token allocations are not additive between auctions – new allocations invalidate old ones.

4.1 Distributed Auctions

In practice, it would be neither practical nor desirable to run central auctions with the participation of all Allocators and Quantifiers in the system. For this reason it is desirable to adopt an approach based on deployment of multiple auctions, each with limited attendance. The can be achieved by letting each Quantifier run its own auction and having Allocators divide their bids in L parts (where $L < I$) and submit the parts to different Quantifiers. The bid from Allocator k to Quantifier i is then written as:

$$q_{k,i}(1),\ldots,q_{k,i}(J) = a_{k,i}(1)q_k(1),\ldots,a_{k,i}(J)q_k(J) \qquad (4)$$

where $a_{k,i}(j)$ are weights which determine how much of the demand at SSP k with respect to service class j is directed to SCP i. It is required that all weights are non negative and for all j and k sum up to one over all i. The simplest approach to setting the values of $a_{k,i}(j)$ would be to set them to yield equal bids to all SCPs, however as will be described in §6 more advanced, dynamic mechanisms for setting the weights, which utilise information regarding the current state of the network, are possible.

5 Multi-Agent System for 'Ant Colony Optimisation' Control

The multi-agent system that will be described in this section takes its inspiration from previous studies on Ant Colony Optimisation, which is the application of approaches based on the behaviour of real ant colonies to optimisation problems. The approach has been applied to problems such as routing in circuit- [2] and packet-switched [8] networks. These studies have demonstrated that ant-based distributed control can have many advantages, for example: high performance in terms of maximising call throughput; scalability; and adaptability to different network topologies, traffic patterns and transient overloads.

5.1 Background Information

Individual ants are very unsophisticated insects from a behavioural point of view, however colonies of ants, acting as a collective, are capable of performing relatively complex tasks, such as building and protecting their nest, forming bridges, colony emigration and finding the shortest routes from the nest to a food source. Such behaviour arises through indirect communication between the ants, affected through modifications induced in the environment – a process called 'stigmergy'. An example of stigmergy is the means foraging ants use to quickly find the shortest path to a food source: individual ants following a particular path will deposit a type of highly volatile hormone, called a 'pheromone,' giving rise to a 'pheromone trail' for that path. In general, ants faced with a decision between alternative paths will follow the path for which the pheromone trail is strongest. However, the shortest path will tend to acquire stronger trails more rapidly (since the ant's round trip time will be shorter and thus more pheromone will be deposited), thereby causing more ants to follow it. This results in positive feedback, ensuring that the shortest path is quickly identified and utilised by the majority of the foraging ants.

The trail laying behaviour described above can be quite straightforwardly applied to routing in telecommunications networks by replacing the routing tables in network nodes by tables of probabilities representing the 'pheromone strength' for onward routes from that node. Mobile *Ant* agents may then travel between various destinations in the network (selected on the basis of pheromone values) and use information they collect (for example network traversal delay) to update these pheromone tables. In order to route a call or data packet the pheromone table for its destination can then be consulted and the route with the current highest pheromone value selected.

5.2 Multi-Agent System Operation

We will now describe the main points of the operation of an Ant-based IN Load Control strategy. (This operation of the strategy is also illustrated by Fig. 3 below.)

- At intervals of length T_{ant}, for every service type a mobile Ant agent is generated at every one of the SSPs in the network and sent to a selected SCP. Ant generation may be controlled by another agent – an *Ant Controller*, which may vary the generation interval based on its own view of the rate of change of load conditions in the network;
- At each SSP a pheromone table for each service type is maintained. The pheromone tables contain entries for all the SCPs in the network – the entries are the normalised probabilities, P_i, of choosing SCP_i as the destination for a future Ant agent;
- The destination SCP of an Ant is selected using the information in the pheromone table following one of two schemes: either the 'normal scheme' or the 'exploration scheme.' The scheme used is selected at random, but with the probability of using the exploration scheme being much less than that for the normal scheme,
  - In the normal scheme the SCP is selected randomly, the probability of picking SCP_i being the probability P_i indicated in the pheromone table (this is analogous

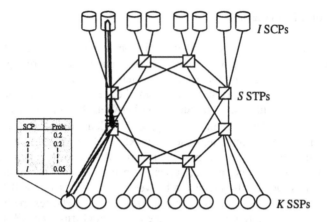

Fig. 3. Ant Colony Load Control

to the trail selection behaviour of real ants);

- In the exploration scheme the SCP is also selected randomly, however the probabilities of selecting all the SCPs are equal. The purpose of the exploration scheme is to introduce an element of noise into the system so that more performant SCPs can be found – this is analogous to a small number of real ants which do not follow the 'recommended' trail, but which, as a result, may find even better trails;

• Ants migrate to the designated SCP, where they interact with the local system and then return to their originating SSP. In doing so they keep track of the time they have spent traversing the network (in both directions);

• Ants arriving at the SCP request information from the local system on the currently expected average processing times for the service type of interest. Processing times reported will be the processing time for the initial message of the service session, and the sum of the processing times for all other messages. It is noted that the separation between the processing times for the initial and subsequent messages is made to highlight the importance of the time spent processing the initial message, during which time the service user will not yet received any response from the network. Reported processing times include those incurred in accessing information from databases, which may be held in SDPs in other parts of the network;

• Upon return to the SSP the Ant uses its gathered information to update the pheromone table entries for its service type. It uses using a formula of the form:

$$P_i = \frac{P_i + \Delta P}{1 + \Delta P}$$

$$P_j = \frac{P_j}{1 + \Delta P}, j \in [1, I], j \neq 1 \tag{5}$$

where i indicates the visited SCP and:

$$\Delta p = \frac{a}{t_1} + \frac{b}{t_2} + \frac{c}{t_3} + \frac{d}{t_4} + e \tag{6}$$

a, b, c, d, and e are constants and

t_1 = time elapsed travelling SSP→SCP

t_2 = expected mean SCP processing for initial message

t_3 = expected mean SCP processing time for subsequent messages

t_4 = time elapsed travelling SCP→SSP

The values of a, b, c and d represent the relative importance given to each of the four measurements;

- Requests for a service are routed to the SCP that has the current highest probability value in the service's pheromone table. It is to be noted that in normal load conditions the operation of the strategy will mean that SCPs with closer proximity to a source are more likely to be chosen as the destination for service requests. This is a result of the fact that delays Ants experience in travelling to/from near SCPs is less than for far ones.

6   Co-operation between Market/Ant Multi-Agent Systems

The Market/Ant MASs described above adopt very divergent approaches towards IN load control. Table 1 below gives an indication of the relevant merits of both approaches based on their ability to meet the IN load control goals outlined in §3. (Note that these evaluations are partially based on the results of the simulation study reported in [5].)

From Table 1 it is clear that a Co-operative Market MAS will be more likely to meet the majority of the goals for IN load control. However the Ant Colony MAS has the advantage that it will balance load in the SS.7 network and takes into account the proximity of SSPs and SCPs. As shown in [9] and [10] overloads of the SS.7 resources can have a devastating effect on the performance of an IN; thus it is desirable that an IN load control strategy incorporates measures to control traffic levels in the SS.7 networks as well as the IN resources themselves. For this reason it is attractive to investigate how both MASs could co-operate to realise an overall agent-based strategy which controls load of both IN and SS.7 resources.

The mechanism we propose for supporting co-operation between the two MASs is to allow Allocators have access to the Ant pheromone values. We envisage two ways in which Allocators could make beneficial usage of the pheromone values:

1. Allocators can use the pheromone values to control the order in which SSPs use the tokens allocated to them. It is desirable to use those tokens associated with SCPs with highest pheromone values first. These SCPs are the ones currently most favoured for that service type – because of their performance, proximity to the SSP, or the state of the network locale the messages traverse to reach them. Note that to operate in this manner it would be necessary to have $T_{ant} \ll T_{auct}$. This approach will be particularly useful in periods of normal traffic loading (when the

Table 1. Qualitative Comparison of operation of Market/Ant MASs

	Co-operative Market MAS	Ant Colony MAS
Goal 1. (Staying within SCP load target)	*YES* Use of tokens limits load offered to the SCP	*NO* No explicit control of traffic volume sent to individual SCPs
Goal 2. (Maximisation of Network Profit)	*YES* Auction algorithm designed to maximise profit (but in the distributed version profit maximisation is not guaranteed)	*NO* No concept of profit values associated with service classes
Goal 3. (Minimisation of response delays)	*YES* Limits on volume of traffic admitted to the network	*NO* Admission of large traffic volumes could cause congestion
Goal 4. (Fair treatment of services/users)	*YES* Services treated fairly but subject to profit related priorities. All users have equal chance of successful service sessions	*YES* No priorities between service types. Users physical location may affect chances of successful service sessions
Goal 5. (Load Balancing between SCPs)	*YES* Tokens allocated for each SCP and all equally likely to be used	*YES* Emergent behaviour of Ants ensures that load balanced in network as a whole
Goal 6. (Load Balancing in SS.7 network)	*NO* No concept of delays in SS.7 or of relative proximity SSPs/SCPs	*YES* Emergent behaviour of Ants ensures that load balanced in network as a whole

majority of allocated tokens will not be used) as it ensures that the 'best' SCPs are given priority;

2. In a distributed auction scenario Allocators can use the pheromone values as the weight values $a_k(j)$, which dictate the amount of expected demand is divided between bids destined for different Quantifier auctions. Basing these bids on pheromone values means that the Allocator is more likely to receive tokens for the SCPs which are currently most favoured for that service type. It is also a dynamic scheme for deciding the contents of bids that will adapt to changes in the traffic levels in the network.

A more practical issue is the means by which pheromone values are passed between Allocators and Ants. One mechanism would be for returning Ants to pass their gather-

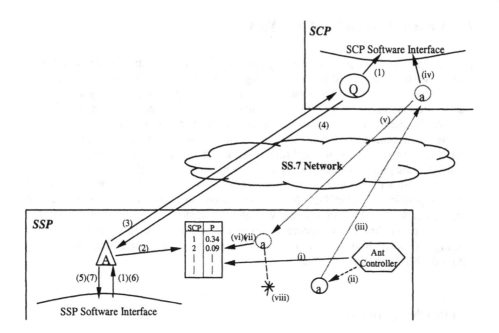

Fig. 4. Co-operation between Market/Ant MASs

ed data to the local Allocator and have the Allocator update and maintain the pheromone table. We contend however that a better solution would be to keep the ph eromone table separate from both Allocators and Ants – this would mean that both types of agents would not need to be aware of the existence of the other. Fig. 4 above illustrates how pheromone values can be used in this manner. The sequence of actions illustrated by Fig. 4 are as follows:

(1) Allocators/Quantifiers collect information for bids from SSPs/SCPs;
(2) Allocator reads current pheromone values;
(3) Allocator sends bids to Quantifiers using pheromone values to divide demands between different Quantifiers;
(4) Quantifier runs auction and returns token allocations;
(5) Allocator specifies which tokens SSP uses first;
(6) SSP asks allocator for new token information – either periodically or when it has used all of previously specified token pool;
(7) Allocator gives SSP new token information (it is noted that the Allocator may consult the pheromone table once more in order to decide which tokens to use next.)

In parallel with the actions of the Market MAS Ant agents are being periodically generated by the Ant Controller based on pheromone values (i)(ii), migrate to a chosen SCP (iii), collect information from the SCP (iv), return (v), read and update pheromone table (vi)(vii) and terminate (viii).

7 Conclusions and Future Work

In this paper we have presented an agent-based load control strategy for Intelligent Networks which is realised by co-operating Market/Ant based MASs. It provides an example of how separate agent systems, each of which realises different goals, can be made to co-operate in a very efficient fashion and with very beneficial results. In our example the agents systems communicate indirectly via a pheromone table. Use of the external environment for communication has the advantages of being highly efficient (which is of importance with a real-time constrained application like load control) and simplifying the process of design and implementation of the MASs.

Future work will concentrate on a quantitative analysis of the likely performance of the co-operating Market/Ant MASs described in the paper. Some further enhancements to both agent systems will also be made. For example in the Market MAS the possibility of Allocators forming a cartel to make a joint bid will be investigated, while in the Ant strategy the possibility of Ants controlling the routing within the SS.7 network will be addressed.

Acknowledgements

This work was supported by the European Commission, through the ACTS project MARINER; the authors wish to acknowledge the valuable contribution of their colleagues in this project to the work.

References

1. Clearwater S. H. (ed.): Market-Based Control. World Scientific, 1996.
2. Di Caro G. and Dorigo M.: Mobile Agents for Adaptive Routing. Proceedings of the 31st Hawaii International Conference on Systems, Big Island of Hawaii, January 6-9, 1998.
3. Harju J., Karttunen T., Martikainen O.: Intelligent Networks. Chapman & Hall, 1995.
4. Lodge F., Botvich D., Curran T.: Using Revenue Optimisation for the Maximisation of Intelligent Network Performance. Proc. 16th International Teletraffic Congress, Edinburgh, June 1999.
5. Arvidsson Å., Jennings B., Angelin L., Svensson M.: On the use of Agent Technology for Load Control, with an example IN 'Market-Based' Strategy. Proc. 16th International Teletraffic Congress, Edinburgh, June 1999.
6. Gibney M., Jennings N.: Market Based Multi-Agent Systems for ATM network Management. Proc. 4th Communications Network Symposium, Manchester, 1997.
7. Ygge F.: Market-Oriented Programming and its Application to Power Load Management. University of Lund, Sweden, Ph.D. Thesis, 1998.
8. Schoonderwoerd R., Holland O. E., Bruten J. L.: Ant-like agents for load balancing in telecommunications networks. Proc. 1st International Conference on Autonomous Agents, 1997.
9. Jennings B., Lodge F., Curran T.: A Strategy for the Resolution of Signalling System No. 7 (SS7) and Intelligent Network (IN) Congestion Control Conflicts. Proc. IEEE International Conference on Communications (ICC), Atlanta, 1998.

10. Atai A. H., Northcote B. S., AIN Focused Overloads - A review of USA CCS network failures and lessons learned. Proc. ITC Mini-Seminar on Engineering and Congestion Control in Intelligent Networks, Melbourne, April 1996.

JAMES: A Platform of Mobile Agents for the Management of Telecommunication Networks

Luis Moura Silva[1], Paulo Simões[1], Guilherme Soares[1], Paulo Martins[1], Victor Batista[1], Carlos Renato[2], Leonor Almeida[2], and Norbert Stohr[2]

[1] CISUC - Dep. Eng. Informatica, University of Coimbra
P-3030 Coimbra, Portugal
luis@dei.uc.pt
[2] Siemens S.A., Rua Irmãos Siemens, N. 1
P-2720-093 Amadora, Portugal

Abstract. This paper presents an overview of JAMES, a Java-based platform of mobile agents that is mainly oriented for the management of data and telecommunication networks. This platform has been developed on behalf of a Eureka Project (Σ!1921) and the project partners are Siemens SA, University of Coimbra and Siemens AG. We describe the main architecture of the platform giving more emphasis to the most important features. To show the effectiveness of some of the techniques that have been implemented we will present some performance results that compare the JAMES platform with the Aglets Workbench.

The main target of our platform is network management and telecommunication applications. In this line, we have done a Java-based implementation of SNMP that has been integrated within the platform. The industrial partners of our project (i.e. Siemens S.A.) have developed a prototype application for TMN performance management. Although it is still a prototype it is being used to validate the technological advantages of using mobile agents in the management of telecommunication networks.

1 Introduction

The main goal of the JAMES project is to develop an infrastructure of Mobile Agents with enhanced support for network management and try to exploit the use of this new technology in some telecommunications software products. A Mobile Agent corresponds to a small program that is able to migrate to some remote machine, where it executes some function or collects some relevant data and then migrates to other machines in order to accomplish another task. The basic idea of this paradigm is to distribute the processing throughout the network: that is, send the code to the data instead of bringing the data to the code.

The existing applications in the management of telecommunication networks are usually based on static and centralized client/server solutions, where every element of the network sends all the data to a central location that executes the whole processing over that data and provides the interface to the user operator. By consequence, they are not flexible, they have problems of scalability and they produce too much traffic in the network.

The use of Mobile Agents in this kind of applications represents a novel approach and potentially solves most of the problems that exist in centralized client/server solutions. The applications can be more scalable, more robust, can be easily upgraded or customized and can reduce the traffic in the network.

The JAMES project will try to exploit all these technological advantages and see how the mobile agents technology can be used in software products that are been developed by Siemens SA. The project involves the development of a Java-based software infrastructure for the execution of mobile agents. The use of Java was motivated for reasons of code portability.

In the last few years the use of Mobile Agent technology has received an extraordinary attention from several Universities and research institutes and a notable investment from several companies [1]. Mobile agents have been applied in several areas, like mobile computing, electronic commerce, Internet applications, information retrieval, workflow and cooperative work, network management and telecommunications [2–5]. Several commercial implementations of mobile agents have been presented in the market, including Aglets from IBM [6], Concordia from Mitsubishi [7], Odyssey from General Magic [8], Voyager from ObjectSpace [9] and Jumping Beans from AdAstra [10]. Although these software products have some very interesting features they are too much general-purpose and do not provide any special support for network management.

In our project, we are developing from scratch a new Mobile Agent infrastructure that is being tuned and customized for the applications we have in mind in the area of telecommunications and data network management. For this reason we decided to develop another platform from scratch that would take into account the following list of issues:

- high-performance and efficient code migration,
- fault-tolerance and robustness,
- support for network management,
- flexible code distribution and easy upgrading,
- mechanisms for resource control,
- disconnected operation,
- easy-to-use programming interface,
- 100% pure Java implementation,
- support for CORBA [11].

These are the main goals of our platform. Some of them have been already achieved in the first release, while the others are scheduled for future versions.

During this project, the platform will be used in two software products: one in the area of telecommunication and other for data network management. Our industrial partners (Siemens S.A.) have been developing a prototype application in the area of performance management by using our platform of mobile agents. This prototype is already finished and we are now conducting a benchmarking study to compare the use of mobile agents over traditional client/server solutions

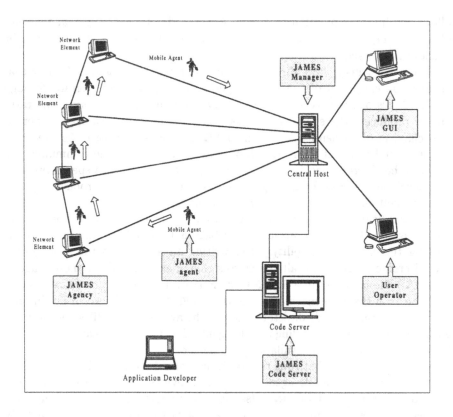

Fig. 1. An Overview of the JAMES Platform

to see if we corroborate some of the advantages of this new paradigm in the field of distributed computing.

The rest of the paper is organized as follows: section 2 presents an overview of the JAMES Platform, giving more emphasis to the most important features of the system. Section 3 presents some experimental results where we compare the performance of JAMES against the Aglets system. Section 4 describes the work we have done in the integration of SNMP in our platform, while section 5 presents a brief overview of the prototype application that has been implemented by Siemens S.A. Section 6 concludes the paper.

2 The Architecture of the JAMES Platform

The JAMES Platform provides the running environment for mobile agents. There is a distinction between the software environment that runs in the manager host and the software that executes in the Network Elements (NEs): the central host executes the JAMES Manager while the nodes in the network run a JAMES Agency. The agents are written by application programmers and will execute on top of that platform. The JAMES system will provide a programming interface that allows the full manipulation of Mobile Agents. Fig. 1 shows a global snapshot of the system, with a special description of a possible scenario where the mobile agents will be used.

Every NE runs a Java Virtual Machine and executes a JAMES Agency that enables the execution of the mobile agents. The agents will migrate through the machines of the network in order to access some data, execute some tasks and to produce reports that will be sent back to the JAMES Manager. There is a mechanism of authentication in the JAMES Agencies to control the execution of agents and to avoid the intrusion of non-official agents. The communication between the different machines is done through stream sockets. A special protocol was developed to transfer the agents across the machines in a robust way and to provide atomicity to the occurrence of failures.

The application developer writes the applications that are based on a set of mobile agents. These applications are written in Java and should use the JAMES API for the control of mobility. After writing an application the programmer should create a JAR with all the classes that make part of the mobile agent. This JAR file is placed in a JAMES Code Server. This server can be a different machine or in the same machine where the JAMES Manager is executing. In both cases, it maintains a code directory with all the available JAR files and provides a mapping to the corresponding mobile agents. The Code Store can be replicated if we want to increase the availability of the code.

The host machine that runs the JAMES manager is responsible for the whole management of the mobile agent system. It provides the interface to the end-user, together with a Graphical User for the remote control and monitoring of agents, places and applications. A snapshot of this interface is presented in Fig. 2. The JAMES GUI is the main tool for management and administration of the platform. With this interface, the user can manage all the Agents and Agencies in the system.

The JAMES platform is still in its first version but the main features of the platform can be summarized in the following list:

- Kernel of the JAMES Manager and JAMES Agency,
- Service for remote updating of Agents and Agencies,
- Flexible code distribution (caching and prefetching schemes),
- Atomic migration protocol,
- Support for fault-tolerance through checkpoint-and-restart,
- Reconfigurable itinerary,
- Support for disconnected computing,
- Watchdog scheme and system monitoring,
- Mechanisms for resource control,
- Logging and profiling of agent activity,
- GUI interface to allow the remote control of agents,
- Interface with CORBA,
- Integration of a Java-based SNMP stack into the platform,
- Inter-agent communication (through JavaSpaces [12]),
- Multi-paradigms for agent execution (simple agent, migratory agents and Master/Worker model).

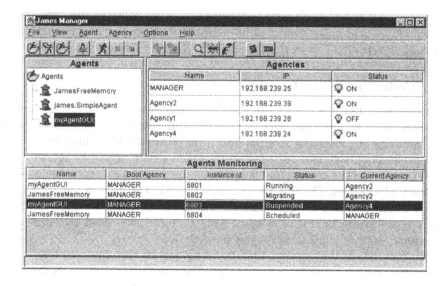

Fig. 2. The Graphical User Interface

The explanation of all these mechanisms is out of scope of this paper, but we will give some emphasis to some of the issues that have been considered as more important for our domain of applications: high-performance, flexible code distribution, remote software upgrading, reliability and robustness.

2.1 Flexible Code Distribution

Each Mobile Agent has a specific Itinerary with a set of tasks (Missions) to be executed across the JAMES Agencies. The agent is usually created and dispatched by the JAMES Manager.

The code classes of the JAMES agents are grouped in JAR files - there is one unique JAR file per agent. The first step is to register the agents in the platform. All the JAR files that have been registered are stored in the Code Store.

To enable a more efficient migration of the agents we have implemented a flexible distribution of their code. The Agencies have their own local disk repository of agent JAR files (disk cache) and there is also a local memory cache of agent classes per Agency. The memory cache is further separated between the different agents running on the same Agency. The disk and the memory cache make use of a LRU policy to replace the JAR files when the cache is full. The cache size can be customized by the platform administrator.

Every time an agent arrives at an Agency there is a hierarchical progressive scheme to fetch the classes of the agent: first, the agent classes are searched in the memory cache; then, in the local disk cache. If the JAR file, associated to the agent, is not in the local cache the Agency tries to get it from the previous Agency of the Itinerary. If the classes could not be fetched from this machine then the Agency contacts the central Code Store.

This hierarchical search scheme provides great flexibility in the code distribution. There is always a central point where all the code is stored but there are multiple choices that can be used to improve performance and scalability.

2.2 Code Prefetching for High-Performance

The caching of JAR files tries to exploit the locality of code of the mobile agents. However, we have improved even further the performance of the agent system by optimizing the inner parts of the migration protocol and by implementing a code prefetching technique to speed up the execution of migratory agents.

When an agent is created and launched to the JAMES platform it has its own Itinerary of Agencies. There is always an overhead in the startup time when the classes of the agent are not in the local caches of the Agencies. This startup time corresponds to the time that is spent to fetch the classes from the local disk repository, or from another Agency or from the Code Store.

We can reduce this stratup time by informing all the Agencies that some particular agent is going to execute there. The JAMES Manager sends some information to all the Agencies (except the first one) of the Itinerary. With this information the remaining Agencies can load in advance the class files while the agent is still executing in the first Agency of the Itinerary.

The JAMES Manager maintains a global information about the caching information of the Agencies and it can even send the JAR file in advance to some Agency, if it that file does not exist in the remote cache.

2.3 Remote Software Upgrading

The JAMES platform provides a flexible mechanism for the remote upgrading of mobile agents, as well as Agencies. Each Agency is seen as a stationary agent: it cannot move around the network once installed in a machine, but it should be easy to upgrade, customize and install by a central host.

The JAMES Agency is composed by two modules: a small `jrexec` daemon and the Agency itself. The `jrexec` daemon is a static piece of software; once installed it does not need to be constantly upgraded. The Agency itself is a more dynamic module, since it can be changed whenever required.

The Java `jrexec` daemon implements an instance of the *Class Loader* and receives some network commands regarding the installation and control of the JAMES Agency. This daemon will be instantiated every time the machine is booted. The daemon can receive a JAR file containing a JAMES Agency and it will perform its local installation. After this first step, the JAMES Manager can send some remote commands to the `jrexec` Daemon:

- Refresh the local memory by calling the Java garbage collector;
- Kill the local Agency;
- Install a new Agency on the local machine;
- Upgrade the local Agency with a new set of classes.

This scheme will be useful in dynamic environments since it provides a flexible way to upgrade remote software.

2.4 Fault-Tolerance

The JAMES platform has some special support for fault-tolerance. The first version includes a checkpoint-and-restart mechanism, a failure detection scheme, an atomic migration protocol and some support for fault-management. The platform should be able to tolerate any failure of a mobile agent, a JAMES Agency or the JAMES Manager.

Fault-Tolerance at the Agencies. Periodically, the internal state of the JAMES Manager and Agencies are saved as a checkpoint in persistent storage. The internal state consists of all the internal objects that keep all the relevant state about the platform and the execution of the agents. If any of the servers (Agency or Manager) fails or is simply shut down the system should have enough information to recover the server to a previous consistent state. This state is retrieved from persistent storage and all the internal state can be reconstructed. The checkpointing mechanism makes use of the Java object serialization facility and is completely transparent to the application programmer.

Fault-Tolerance at the Mobile Agents. If there is a communication or node failure that affects the execution of the agent the system should have some way to assure a forward progress of the mobile agent. This is also achieved through a checkpointing mechanism. When a mobile agent finishes a task in a JAMES Agency its internal state is saved to stable storage before being transmitted to the next destination. The agent is migrated to another host but its data will remain in stable storage until it has been successfully restarted in the next Agency. When it is restarted in the new place the system takes a new checkpoint of the agent and the previous place is informed. The previous checkpoint of the agent can then be removed from stable storage. This checkpointing mechanism is transparent to the application developer and is incorporated in the migration protocol to assure the atomicity of the agent transfer. This means that or the agent is completely migrated to its destination or whenever is a failure the agent is not lost and the system is able to recover the agent in the previous Agency. We have used a conventional two-phase commit protocol to achieve the exactly-once property in the migration of the agents.

When there is a failure in the migration protocol or if one of the Agencies in the Itinerary is not available the agent can execute one of the three following procedures:

1. go back to the JAMES Manager;
2. jump to the next available Agency in the itinerary;
3. or just wait until the destination Agency is back alive.

The procedure to follow by a mobile agent in the occurrence of a failure can be customized by the application programmer.

Resource Control. One important feature in a platform of mobile agents is a good set of mechanisms for resource control. In the JAMES platform we have included some schemes to control the use of some important resources of the underlying operating system, namely: the use of threads, sockets, memory, disk space and CPU load.

These schemes have proved to be very effective when we were doing some stress testing. In some situations when the Agencies are running almost out of any of those resources it was still possible to maintain the platform up and running. Without such mechanisms the Agencies would normally hang up. With resource control the platform has become much more robust and this is a crucial step if we want to use it in production codes.

3 Some Performance Results

In this small section we would like to show the effectiveness of some of the techniques that we have implemented in the JAMES platform.

We decided to compare the performance of our platform with the Aglets Workbench from IBM Tokyo [6]. We used a simple mobile agent that roams a network of five computers to get a report about the free memory of each machine. These machines were running Windows NT and JDK 1.1.5 and were connected through a dedicated 10 Mbit/sec Ethernet network. Each machine was executing a JAMES Agency, while the JAMES Manager was installed in a separated computer.

In Fig. 3 we present the results that were taken when executing by the first time the mobile agent with the JAMES platform and with the Aglets system. The results were taken with two versions of JAMES: one that makes use of the prefetching techniques (JAMES-Pref) and another one that uses the normal code distribution procedure. We increased the internal size of the agent to see its relevance in the migration time. We used three different sizes: 1 Kb, 100 kb and 1 Mbytes.

In both cases the JAMES platform presented a better performance than the Aglets system. When using the prefetching technique we were able to improve the performance against the normal version of JAMES by a factor of 1.8. Also, the JAMES platform with the prefetching technique was 4 times faster than the Aglets system. It can also be seen that the performance gap between JAMES and Aglets increases directly when we augment the size of the mobile agent.

JAMES is being developed with high-performance in mind since this is an important issue in our target domain of applications while the Aglets Workbench is mainly oriented for Enterprise computing and Internet applications. In these two areas performance is not a so crucial aspect.

In the previous experiment we were executing the mobile agent by the first time in the network. The experiment was useful to measure the importance of the prefetching technique but the application did not make any use of the caching scheme. In the next experiment, whose results are presented in Fig. 4, we present the second execution of the mobile agent using JAMES and Aglets. This time the class files of the agent were residing in the local caches of the Agencies in the

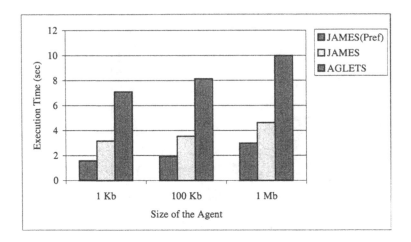

Fig. 3. Comparing the performance of JAMES with Aglets Workbench (the use of prefetching)

network. The JAMES platform was in average 4.3 times faster than the Aglets system for the second execution of the agent. The interesting aspect is the fact that Aglets also has a caching mechanism inside its platform, but apparently it was not so efficient as the scheme implemented on JAMES.

Although this section did not present a comprehensive study, the performance results of the JAMES platform seem to be quite promising when compared with a general-purpose system of mobile agents.

4 Integration of SNMP

In this section we describe in some detail the support we have included in JAMES platform for SNMP-based network management.

4.1 SNMP and Mobile Agents

Classic management applications use protocols like SNMP [13] and CMIP [14] to interface with management services in heterogeneous environments. Mobile Agents will not replace these protocols. Instead, they will complement them with powerful programming metaphors allowing more efficient solutions for network management.

Mobile agents provide a very attractive approach to incorporate mobile code into the existing local management services, in order to perform intelligent tasks closer to management data [15]. However, some management protocols are still necessary to retrieve and process the management information. This is the case when some of the managed NEs are unable to host mobile agents, the management services of NEs are not directly available to the hosted mobile agents or the interfaces with management services are non-standardised.

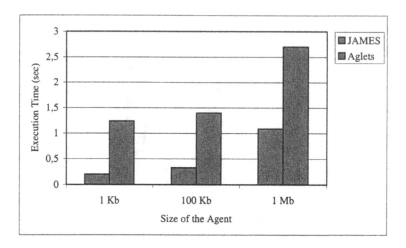

Fig. 4. Comparing the performance of JAMES with Aglets Workbench (the use of caching)

Integration of management protocols with mobile agents can be relegated to the applications' developer, eventually using the same "general-purpose" libraries used by static management systems. However, code mobility, security constraints and resource usage control imposed on mobile agents' applications seriously limit the usage of these protocols without explicit support from the underlying infrastructure. This is why JAMES includes explicit support for interoperability with SNMP devices and applications.

There are now several worthwhile projects mixing SNMP with mobile agents, like the Perpetuum Mobile Procura Project [16, 17], the Discovery platform [18], the Astrolog/Magenta platform [19] and the INCA Architecture [20]. Some of these projects relegate SNMP support to applications' developers whilst others integrate SNMP within the mobile agents' infrastructure. The JAMES platform also integrates SNMP within the platform, but departs from these projects for the following reasons:

- it provides maximum interoperability, since it covers three different service ranges: interaction with local and remote SNMP-agents; interaction between SNMP-managers and mobile agents; and infrastructure management using SNMP;
- it provides full support for agents mobility;
- it is a completely optional feature of JAMES, imposing no additional overheads when turned-off;
- it does not require any intervention on existing SNMP devices and SNMP Managers, preserving the overall portability.

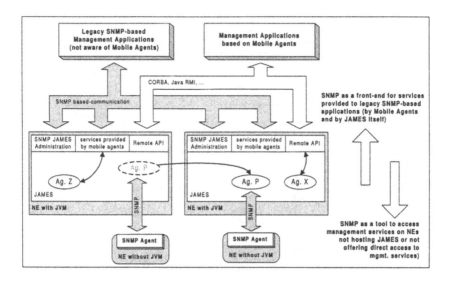

Fig. 5. Proposed Integration Framework

4.2 SNMP Services Provided by JAMES

JAMES includes a framework of full-fledged SNMP services already integrated and available to the NM-application developer, resulting in broader application fields and reduced development costs. As represented in Fig. 5, three nuclear SNMP services were considered:

- a service allowing mobile agents to interact with SNMP-agents, acting as SNMP-managers;
- support for communication between SNMP-managers and mobile (or stationary[1]) agents;
- a management service allowing legacy management platforms to administer the JAMES infrastructure itself using SNMP.

These services provide the following features:

- management of NEs not supporting JAMES Agencies but equipped with SNMP-agents;
- management of NEs supporting JAMES Agencies but restricting direct access to management information for security or architecture reasons;
- management of the JAMES infrastructure itself as an SNMP-service;
- usage of mobile agents to deploy intermediary management services layered between NEs (SNMP capable or not) and legacy SNMP-managers. These could be new services or just management information processing closer to the NEs;
- usage of mobile or stationary agents for fast development and deployment of SNMP services.

---

[1] JAMES supports "stationary agents" in the sense that agents can make little or no use at all of their migration capability.

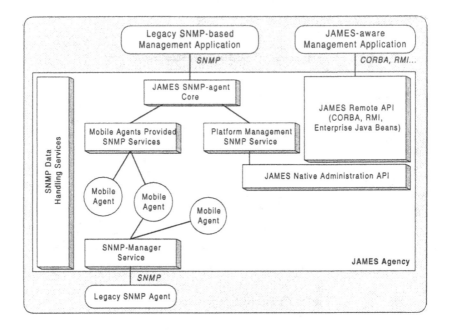

Fig. 6. High-Level Structure of SNMP Services of JAMES

4.3 Design of SNMP Services

Since SNMP is just one of several protocols to be used by network management applications, the following design constraints were defined:

- SNMP support must be optional, not increasing resource usage when turned-off;
- SNMP support should not increase the complexity of the platform;
- design of SNMP support may not compromise the platform scalability and functionality;
- SNMP support must be portable across different hosts without conflicting with SNMP services already installed in the hosts, like native SNMP agents.

These resulted in a modular design Fig. 6 where SNMP services are placed outside the platform core and can be dynamically installed and removed, without imposing a permanent overhead in the JAMES infrastructure. Most services consist themselves of mobile agents (the "Service Agents", granted with exceptional permissions to access necessary resources) providing services to common agents through inter-agent communication. The Agency offers a directory service where common mobile agents can locate the Service Agents (or implicitly require their installation). This solution provides an elegant lightweight framework to support specific services. In the future, new kinds of services can be easily integrated into the JAMES platform.

SNMP Data Handling Services. These services include all the tools needed to handle SNMP data and SNMP Protocol Data Units. These tools are available as a set of Java classes for high level representation of SNMP data types and PDUs, and for ASN.1/BER encoding [21]. Mobile agents impose no particular requirements to this Service, meaning general purpose Java-based SNMP tools, like [22], could have been used without prior adaptation.

SNMP Manager Service. This service allows mobile agents to interact with SNMP Agents using a manager-API, to query SNMP-agents, and a Trap Listener that receives SNMP Traps and redirects them to the interested mobile agents. When compared with similar Services integrated in classical management applications, this Service presents two key differences: support for mobility - mobile agents receive SNMP Traps independently of their present location and migrate without abandoning ongoing SNMP queries - and the service location within the platform - based on the already mentioned "Service Agents".

The SNMP Manager-API is based on the traditional concepts in most high-level SNMP stacks (*sessions* or *contexts*, *request* operations and event handlers), with protocol details being handled in a transparent way. This Service, located in "Service Agents", might be replaced with a third-party "classic" SNMP stack integrated in the Agent's code[2], trading-off mobility support.

The Trap Listener also uses traditional concepts found on other Trap Multiplexers. Mobile agents register their interest on the reception of certain SNMP Traps. Registrations may be valid just while the agent remains at the Agency, for a pre-defined period of time or for the agent's entire lifetime (in the last two options, the Trap Listener may have to forward the Trap Arrival Notification to the new location of the agent). The arriving SNMP Traps produce Trap Arrival Notifications.

Services for Interoperability with Legacy SNMP Managers. While the Service described in the previous section covers communication between mobile agents and SNMP agents, there three other services providing interoperability with legacy SNMP Managers (Fig. 7).

The SNMP-agent Core maintains an SNMP-agent with data being supplied by the two underlying services. The present interface used to register new variables or groups at the MIB-table, to issue Traps and to reply to SNMP requests is proprietary, although future use of standard protocols for expansible SNMP agents like DPI [23], as proposed by [24] is not excluded.

The JAMES SNMP-agent is independent of eventually existing native SNMP Agents, although integration would be more in the spirit of SNMP. Since native

---

[2] This is based assumption that since JAMES mobile agents implicitly control their migration, they can delay migration whenever completion of on-going SNMP transactions is crucial. This degrades performance and affects the programming model (agent migration has to become aware of SNMP transactions) but still allows for some mobility.

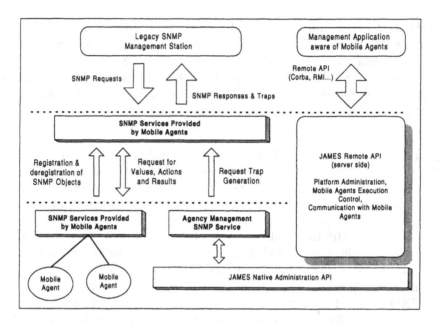

Fig. 7. JAMES Services for Integration with Management Applications

agents are either monolithic or based on a wide diversity of agent-expansion mechanisms, like SMUX [25] or DPI, there is no truly portable and non-intrusive integration method.

The JAMES SNMP-agent allows SNMP communication between legacy applications and mobile agents, opening the way for easy installation of new management services (corresponding to one or several mobile agents) available to legacy SNMP-based network management systems.

In such a scenario mobile agents can be used to pre-process data gathered from existing management services (thus offering higher level functionality), operate as SNMP proxies for NEs using proprietary management interfaces and dynamically install new management services.

The use of stationary or mobile agents for fast deployment of management services available to legacy applications is not new. The JDMK toolkit [26], although not based on mobile agents, shares the vision of a fast and flexible scheme to develop and deploy management services. It includes a complete set of tools to create and remotely install these services using Java and push/pull techniques. The Perpetuum Mobile Procura Project [16, 17] presents a framework where mobile agents use DPI to provide services to SNMP Managers.

The JAMES approach consists of a "service agent" where mobile agents interested on providing an SNMP interface must register their SNMP objects. Later on, SNMP requests from outside applications result in events passed to mobile agents. These events will then trigger predefined management actions resulting in SNMP responses.

The JAMES SNMP agent also allows SNMP-based management of the JAMES platform itself. Although a richer interface is available to dedicated applications using Java RMI and CORBA, a subset of management functions has been "translated" into an SNMP MIB that provides monitoring, fault-management and performance management. This is implemented using a "Service Agent" (the Agency SNMP Management Service) that acts as a translator between the SNMP-agent Core and the internal JAMES administration API (Figure 7). The intention is not to use SNMP to fully administer JAMES but to provide a simple SNMP-based management service.

5 Prototype Application for Performance Management

The software for TMN Performance Management is currently dominated by systems based on Client/Server technologies. In most cases, this approach results in monolithic, not scalable and hardly flexible solutions.

Typically, there are several network devices distributed across the network. In each device it is installed a static server (or agent) that is responsible for collecting raw data the local device, with little or even no pre-processing at all. At the manager host there is a client process that interacts with all the servers in the network devices, collects the information from them and provides the information required to the end-user. These servers (or agents) are static and proprietary processes. They are very difficult to upgrade or to customize. Some of the current solutions also suffer from an information bottleneck at the manager host due to its centralized nature. Some of the applications are very inefficient in the use of the network bandwidth, since most of the data has to be sent from the network devices to the manager site in order to be processed.

Additionally, raw data information generated in the TMN Network Elements has some "troublesome" characteristics, like:

- proprietary raw data formats;
- huge amount of raw data produced at a daily basis;
- variation of data formats depending on customer and/or software version installed.

Given this scenario, the use of Mobile Agents for TMN data collection provides several potential benefits. First of all, mobile agents allow some processing at the data sources which increases the scalability and reduces the traffic in the network. The robustness of the application is also improved through the use of autonomous mobile agents and the application provider will have a more flexible paradigm to deal with the diversity of network configurations. The application developer can create diverse collector agent versions, each one matching the proprietary raw data format involved, the software versions involved and the customer involved. Software upgrading is another important issue that comes from the use of mobile agents. The upgrade of an application requires only the coding of new agents or the automatic deployment of new agent versions over the managed network. This way, the extension of the functionality or the installation of new software versions becomes more flexible and effortless.

In order to evaluate the use of mobile agents and the JAMES platform in real PM/TMN environments we have designed a prototype application to provide O&M Destinations Reporting, a component of TMN Performance Management Traffic measurements [27].

A commercial TMN application from Siemens S.A. already addresses this problem, making it possible to evaluate the advantages/disadvantages of Mobile Agents versus the use of the client/server paradigm in a real TMN environment.

5.1 Overview of the Application

The TMN application from Siemens collects and processes performance data from the NEs in order to produce a set of global reports about the performance of the network. This can be applied to the mobile and switched network. Right now, data collection and report building are two dissociated tasks: data is collected through file-transfer and is organized in a centralized relational database. This database is later queried to produce the performance reports. This two-step approach is based on a traditional client/server approach and presents some technical limitations: the data collection introduces too much traffic in the network and the overall system has some problems of scalability.

The benchmark application, designated as "EWSD Destination Reporting Application", is designed to reproduce a representative subset of the TMN application reports using mobile agents to collect and process the management data. This new version of the application is supposed to overcome some of the problems of the existing client/server version. It is structured into three different modules:

- an application GUI that handles the end-user requests and outputs the reports;
- a mobile agents handler, responsible for the control of the reporting features;
- the application specific mobile agents which are assigned to fulfil the required reports.

The benchmark application produces two different types of reports: *on-demand* and *predefined/scheduled* reports. *On-demand reports* correspond to requests to be immediately executed over the traffic destination raw data files stored in the remote NEs. Fig. 8 presents a snapshot from an *on-demand* report that shows the evolution of two combined traffic destination counters.

Predefined reports are a sort of report *templates* with a pre-defined behavior. This report templates can be used at a later time with user provided attributes like time window to be evaluated, NEs to be considered, type of traffic destination to be analyzed, etc.

5.2 Benchmark Study

The development of the benchmark application is already complete and a benchmarking study is being conducted, focusing on the following metrics:

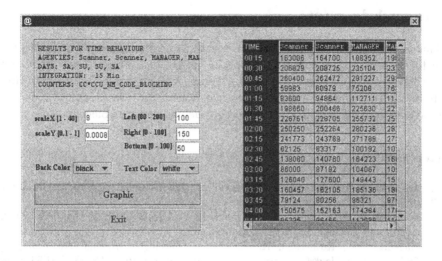

Fig. 8. Snapshot from the Prototype Application: a Time Behavior Report in Raw Table Format

- scalability and traffic bottlenecks;
- dynamic execution of mobile agents;
- performance and mobility;
- network traffic;
- persistency of the applications;
- robustness under stress testing;
- upgrading mechanisms for both agents and agencies;
- easiness to use and develop applications using the JAMES API;
- flexibility of customization.

The complete results of this benchmarking study should be complete within a couple of months.

6 Conclusions

The JAMES Project exploits the paradigm of mobile agents in the field of management of telecommunication networks: the use of Mobile Agents. We are developing a Java-based platform of mobile agents and we have some important goals in mind, like: high-performance, flexible code distribution, remote software upgrading, reliability, robustness and flexible support for SNMP and network management. The preliminary results about the performance of the platform seem to be quite promising.

The JAMES platform will be used in two software products: one in the area of TMN and another for data network management. Within this project we expect to show that Mobile Agents can overcome some of the problems that exist with traditional Client/Server solutions.

Acknowledgements

This project is partially supported by Agência de Inovacão and was accepted in the Eureka Program (Σ!1921). Special thanks to the rest of the team that is working in the project: Luis Santos, Fernando Bernardino and Rodrigo Reis (from the University of Coimbra), Jochen Reban, Patricia Monteiro and João Boavida (from Siemens S.A.) and Norbert Baumgart (from Siemens A.G.).

References

1. Agent Product and Research Activities. http://www.agent.org/pub/activity.html
2. Intelligent Agents. Communications of the ACM. Vol. 37, No. 7, July 1994
3. Hermans, B.: Intelligent Software Agents on the Internet. http://www.hermans.org/agents/index.html
4. Magedanz, T., Rothermel, K., Krause, S.: Intelligent Agents: An Emerging Technology for Next Generation Telecommunications. Proc. INFOCOM'96, San-Francisco (1996)
5. Pham, V.A., Karmouch, A.: Mobile Software Agents: An Overview. IEEE Communications Magazine, pp. 26-37, July 1998
6. IBM Aglets Workbench, http://www.trl.ibm.co.jp/aglets/
7. Concordia, http://www.meitca.com/HSL/Projects/Concordia/
8. General Magic Odyssey, http://www.genmagic.com/agents/
9. Voyager, http://www.objectspace.com/voyager/
10. Jumping Beans, http://www.JumpingBeans.com/
11. OMG: The Common Object Request Broker Architecture and Specification. (1995)
12. JavaSpaces, http://java.sun.com/products/javaspaces
13. Rose, M.: The Simple Book - An Introduction to Management of TCP/IP-based Internets, 2nd Edition. Prentice-Hall International Inc. (1994)
14. ISO/IEC 9595: Information technology - Open Systems Interconnection - Common management information Service definition. International Organization for Standardization, International Electrotechnical Commission (1990)
15. Goldzmith, G., Yemini, Y.: Decentralizing Control and Intelligence in Network Management. Proceedings of 4th International Symposium on Integrated Network Management, Santa Barbara (1995)
16. Perpetuum Mobile Procura Project, Carlton University, http://www.sce.carleton.ca/netmanage/perpetum.shtml
17. Bieszczad, A.: Advanced Network Management in the Network Management Perpetuum Mobile Procura Project. Technical Report SCE-97-07, Systems and Computer Engineering, Carleton University (1997)
18. Lazar, S., Sidhu, D.: Discovery, A Mobile Agent Framework for Distributed Application Development, Technical Report, Maryland Center for Telecommunications Research, University of Maryland Baltimore County (1997)
19. Sahai, A., Morin, C.: Enabling a Mobile Network manager (MNM) Through Mobile Agents. Proceedings of Mobile Agents, Second International Workshop MA'98, Stuttgart, Germany (1998)
20. Nicklish, Quittek, J., Kind, A., Arao, S.: INCA: an Agent-based Network Control Architecture. Proceedings of IATA'98, Paris (1998)
21. Information Processing, Open Systems Interconnection: Specification of Basic Encoding Rules for Abstract Syntax Notation One. ISO (1987)

22. AdventNet SNMP, http://www.adventnet.com/products/snmpbeans

23. Wijnen, B., Carpenter, G., Curran, K., Sehgal, A., Waters, G.: Simple Network Management Protocol Distributed Protocol Interface Version 2.0, RFC 1592 (1994)

24. Susilo, G., Bieszczad, A. and Pagurek, B.: Infrastructure for Advanced Network Management based on Mobile Code. Proceedings of the IEEE/IFIP Network Operations and Management Symposium NOMS'98, New Orleans (1998)

25. Rose, M.: SNMP MUX protocol and MIB. RFC 1227 (1991)

26. Java Dynamic Management Kit, http//www.sun.com/software/java-dynamic

27. ITU-T Recommendation M.3010: Principles for a Telecommunications Management Network. (1992)

Using Mobile Agents for Distributed Network Performance Management

Damianos Gavalas[1], Dominic Greenwood[2], Mohammed Ghanbari[1], Mike O'Mahony[1]

[1]Communication Networks Research Group,
Electronic Systems Engineering Department,
University of Essex, Colchester, CO4 3SQ, U.K.
E-mail: {dgaval, ghan, mikej}@essex.ac.uk
[2]Distributed Network Management and Agent Technology Research Group
Fujitsu Telecommunications Europe Ltd.,
Northgate House, St. Peters Street, CO1 1HH, Colchester, U.K.
E-mail: D.Greenwood@ftel.co.uk

Abstract. The intrinsic limitations of traditional centralised Network Management (NM), such as information bottlenecks and lack of flexibility, have encouraged a trend towards distributed management intelligence. Although several distributed NM architectures, exploiting the advantages of Mobile Agents (MA) have been recently proposed, when considering Network Performance Management (NPM) they fail to address scalability problems. In this paper, we describe a secure and fault-tolerant management framework based on MAs, which addresses these limitations by introducing two efficient, lightweight polling modes. Both real-time and off-line NM data acquisition is considered. An in-depth performance analysis of the introduced polling modes, in a data-intensive NPM application is also undertaken. The two modes are shown to outperform SNMP-based polling both in terms of response time and bandwidth consumption.

1. Introduction

The current state of the art in NM involves a management application (*manager*) and the managed entities (*agents*[1]), embedded within Network Elements (NEs). Management interactions make use of a centralised Client/Server (C/S) model, with the manager (client) collecting status data and setting control variables through the agents (servers). Communication between the managing and the managed parties is facilitated by NM protocols such as the Simple Network Management Protocol (SNMP) [1], part of the TCP/IP protocol suite and the Common Management Information Protocol (CMIP) [2] used in public telecommunication networks. Within these protocols, physical resources in a network are represented by managed objects.

---

[1] The term 'agent' here refers to static management agents, which should not be confused with Mobile Agents.

Collections of managed objects are grouped into tree-structured Management Information Bases (MIB)[2] following the Abstract Syntax Notation 1 (ASN.1) format.

Network Management Systems (NMS) based on this C/S archetype exhibit several drawbacks: due to rigid design time definitions, NMS functionality cannot be dynamically updated, whilst frequent polling of NEs for management data is known to result in substantial data transmission rates between the manager and agents. This creates a processing bottleneck at the manager host and adds a considerable strain on network throughput [6].

All these problems suggest distribution of management intelligence as a rational approach to overcome the limitations of centralised NM. The IETF has proposed an approach, known as RMON (Remote MONitoring) [4], which introduces a degree of decentralisation. In particular, RMON network monitoring devices (*probes*) collect management statistics from their local domain (e.g. an Ethernet segment), providing detailed information concerning traffic activity. However, this approach is expensive as it typically requires a stand-alone RMON compliant device (probe) in every network segment.

In terms of current research activities, Management by Delegation (MbD) [5] represents a first clear effort towards decentralisation. The idea of management distribution is taken further by solutions that exploit Mobile Agents (MA), which provide a powerful software interaction paradigm that allows code migration between hosts for remote execution. The data throughput problem can be addressed by delegation of authority from managers to MAs, which are able to filter and process data locally without the need for transmission to a central manager. This ability has attracted much attention to MA technology, with several Mobile Agent Frameworks (MAF) proposed for NM applications [7][8][9]. Neither [7] nor [8] though, consider remote processing issues and, as a result, the manager host still suffers from a computational burden induced by processing bottlenecks. These issues are addressed in [9].

It should be emphasised that all the aforementioned works involve frequent MA transfers, when the collection of management statistics is considered. In addition to the apparent network overhead, these frameworks are not scalable as they assume a 'flat' network architecture, i.e. a single MA is launched from the manager platform and sequentially visits all the managed NEs, regardless from the underlying topology. Thus, for large networks the round-trip delay for the MA will greatly increase, whilst the extracted statistics will not be accurate and reliable due to the non-negligible time intervals between the acquisition of each data sample for every NE. For such reasons, these MAFs are not appropriate for Network Performance Management (NPM), which represents the main focus of this paper.

NPM involves gathering and logging data generated by devices, which may be analysed off-line or in real-time. That process helps in measuring the performance, throughput and availability of network resources. The advantage of analysing the data in real-time is that it allows sophisticated NMSs to foresee a possible congestion or failure and take preventive measures before the actual error occurs. On the other hand, collected data may be used to build daily, weekly or monthly graphs/reports to assist the administrator in network planning. In such cases there is no need for real-time NM

---

[2] MIB is an SNMP term, which corresponds to the term Management Information Tree (MIT) used in CMIP. Hereafter, we ignore the difference for sake of simplicity.

data and, hence, an alternative and complimentary polling mechanism should be applied.

Therefore, we propose two MA-based polling modes intended to provide an efficient method for obtaining both real-time and off-line management data. In the first approach, called *Get 'n' Go* (GnG), used to collect real-time data, the network is partitioned into several domains and a single MA object is assigned to each of them. In every Polling Interval (PI), this MA sequentially visits all NEs within the network domain and obtains the requested information before returning to the manager. The second polling scheme, called *Go 'n' Stay* (GnS), targets the acquisition of data to be analysed off-line, where the need to obtain data in short time frames is no longer an imperative. Thus, we introduce a method where an MA object is broadcasted to all managed devices; the MA remains there for a number of PIs and collects an equal number of samples before returning to the manager. The infrastructure described in [9] has been extended to support the introduced polling modes.

The paper is organised as follows: Section 2 provides an overview of the MAF used to support this work, while Section 3 describes in detail the two introduced polling modes. Section 4 discusses how MAs can perform semantic compression of NM data, with a performance analysis given in Section 5 and experimental results reported in Section 6. Section 7 concludes the paper and suggests several topics for further work.

2. Mobile Agent Network Management Framework

The MA-based NM framework has been entirely developed in Java [12] as it offers the platform independence required for the management of distributed heterogeneous environments. This and other features, such the rich class hierarchy for communication in TCP/IP networks and its already wide acceptance for the development of distributed applications, positions Java as an ideal platform for MA-orientated management services. The key assumption of our approach is the presence of Java Virtual Machine (JVM) in every NE that is needed to host MAs as active processes. Current trends [13] indicate that Java-enabled network devices are beginning to emerge into the open marketplace.

Our framework consists of four major components [9], illustrated in Figure 1:
1. The Manager application;
2. The Mobile Agent Server (MAS);
3. The Mobile Agent Generator (MAG);
4. The Mobile Agents.

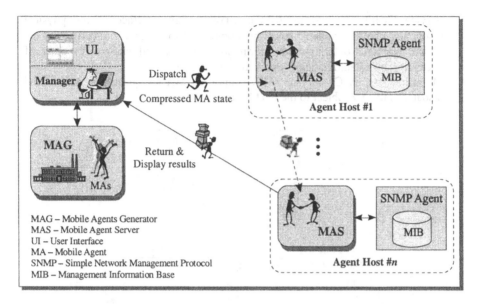

Fig. 1. The Mobile Agents-based Infrastructure

2.1. Manager Application

The manager application, equipped with a browser style User Interface (UI), coordinates monitoring and control policies relating to the NEs. Active agent processes are *discovered* by the manager, which maintains and dynamically updates a 'discovered list'.

2.2. Mobile Agent Server (MAS)

The interface between visiting MAs and legacy management systems is achieved through MAS modules installed on every managed device. The MAS resides logically above the standard SNMP agent, creating an efficient run-time environment for receiving, instantiating, executing, and dispatching MA objects, whilst protecting the system against external attack.

The MAS composes four primary components (see Figure 2):

- Mobile Agent Listener (MAL): a daemon that listens on well-known TCP and UDP ports for incoming MAs. Upon receiving an MA, its state is compressed and *de-serialised* [14].
- Security Component (SC): acts as the system's protective shield. The RSA algorithm [15] has been implemented to provide both *authentication* of incoming MAs and *encryption* of sensitive NM information.

- Service Facility Component (SFC): serves as an interface to the physical resources. The MA makes use of this component to interact indirectly with the SNMP agent and obtain system information. The acquired data are processed, if necessary, by an automatically invoked MA method.
- Migration Facility Component (MFC): upon an MA's request, it serialises and dispatches the MA object to the next host.

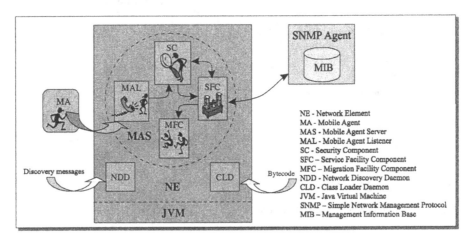

Fig. 2. The Mobile Agent Server structure

Two additional threads are present on each NE outside the boundary of the MAS: the Network Discovery Daemon (NDD) that allows the manager to discover active agent processes and the Class Loader Daemon (CLD), which receives and loads the MAs classes (bytecode).

Enhanced Security Pattern: The security pattern of [9] has been improved by introducing *authorisation* features that restrict the authority domain of visiting MAs on legacy systems. Specifically, the Java standard Security Manager (SM) has been extended to prevent MAs from directly reading/writing files, creating class loaders or sub-processes, exiting the MAS application, etc. This is achieved through registering the incoming MA threads to a given *thread group*, i.e. a batch of threads that eases the manipulation of active MAs. Based on the fact that an MA cannot change the thread group it belongs to, whenever a malicious action is detected, the SM checks the thread group of the action originating thread; when this thread belongs to the MAs thread group, the action is not permitted.

The only problem not addressed by the current implementation of the SM is that MAs cannot be restricted from entering an endless loop (consuming all CPU cycles), or creating thousands of new threads. However, in such an emergency situation, the SM could selectively destroy all threads not belonging to the MAS components' thread group. With these security extensions, we consider the network devices to be relatively safe from malicious MA attacks.

2.3. Mobile Agent Generator (MAG)

The MAG is essentially a factory for constructing customised MAs in response to service requirements. Such MAs may dynamically extend NMS functionality, post MAS initialisation, to accomplish management tasks tailored to the needs of a changing network environment.

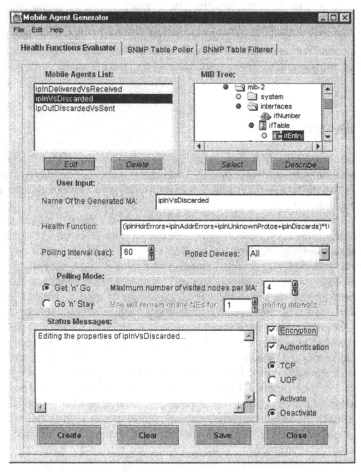

Fig. 3. Mobile Agent Generator User Interface

The MAG's operation is described in detail in [9]. However, its functionality has been extended so as to allow the operator (through the dedicated UI shown in Figure 3) to specify:

- Whether the constructed MA will be used for GnG or GnS polling;
- The polling frequency (i.e. the PI's duration);
- The transmission protocol to be used for the MA transfers (either TCP or UDP);
- Whether MAs authentication and data encryption are applied or not.

It should be emphasised that the transfer of the MA bytecode is performed only *once*, through broadcasting it to the active MASs at MA construction time. Thereafter, the transfer of persistent state, obtained from serialising the MA instance, is sufficient for the MAS entities to recognise the incoming MA and recover its state. This approach relieves the need to transfer MA bytecode to the same NEs during every PI. In contrast, [7] and [8] apply a policy that requires the transfer of both the MA's bytecode and persistent state, resulting in a much higher demand on network resources, as code size is typically much larger than state size [11].

Pre-defined MA property sequences are stored in configuration files and parsed at manager initialisation to instantiate corresponding MA objects. These properties may be modified subsequent to agent instantiation; the modifications will take instant effect.

2.4. Mobile Agent

From our perspective, MAs are Java objects with a unique ID (consisting of the manager host network address and an individual sequence number), capable of migrating between hosts where they are executed as separate threads and perform specific management tasks.

The MA model described in [9] has been refined by introducing an MA *superclass*, which provides some root attributes: an itinerary, a data folder where management information is stored, a problem folder used to report faults. In addition, we have implemented several service-oriented classes that extend the MA superclass. These in turn, are sub-classed by MA classes created by the MAG. Superclass methods are either invoked at MA instantiation, arrival, departure or used to control interaction with polled devices. This flexible hierarchical approach minimises MA bytecode and eases the creation of service specialised MAs.

Also to protect the MAs against tampering, sensitive MA properties (e.g. ID, itinerary, etc.), may be specified only once, when the MA is instantiated. If a malicious host attempts to modify this information, an exception is thrown.

Itinerary, data and problem folders have been implemented as Java *vectors*, i.e. dynamic arrays. At each visited NE, a data sample is appended to the data folder whereas the local NE address is removed from the itinerary vector. In this way and by applying data aggregation methods, as described in Section 4, we prevent the MAs state from growing too rapidly.

The MA's state information is compressed (using the Java *gzip* utility) before being transferred to the next destination host to minimise communication overhead.

2.5. Tolerating Node Failures

An MAF should be able to survive failure scenarios by improving the infrastructure's fault tolerance in terms of securing MA migrations.

Scenario 1: An MA should adapt to unexpected situations, such as the failure of a host MAL thread. In this case, if TCP is used for MA transfers, the TCP connection establishment fails (Figure 4(a)). The MA's *onFailMigration()* method is then automatically invoked to record the unreachable host's name into the MA's problem folder and retrieve the next destination host from the itinerary vector (Figure 4(b)). The MA will then migrate to this host (Figure 4(c)). When returning to the manager host, the MA reports the failed device threads to the manager application, which in turn will take any necessary regenerative actions.

If UDP is the transport protocol choice, the detection of a failed MAL thread is more complex as the UDP datagram carrying the MA's state would simply be lost. A feasible solution to this problem would be to enforce the destination host to issue an acknowledgement when successfully receiving an incoming MA. In case of a fault, the acknowledgement would never reach the originating node and after a given time interval the MA image would be transmitted to its next destination. However, such an approach would create an additional traffic load, thereby reducing the benefit of UDP's lightweight nature.

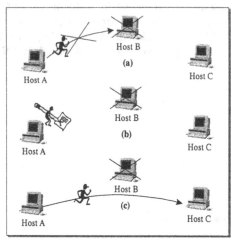

Fig. 4. MA reaction to the detection of a failed MAL thread

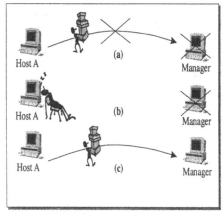

Fig. 5. The specific case of MAL thread failure on the manager host

Scenario 2: A second scenario would be a fault in the MAL thread of the manager host itself (Figure 5(a)), which represents a special case. Since loss of management data carried by an MA should be avoided, an alternative approach to scenario 1 is employed. Upon detecting a fault, the MA will 'sleep' for a given interval (Figure 5(b)) and then resume execution in order to retry the connection. If the manager application recovers before a pre-determined number of retries elapses, the MA is transferred (Figure 5(c)), otherwise it is disposed of locally.

3. Polling Modes

3.1. Get 'n' Go Polling Mode

Traditional SNMP-based polling, which is intrinsically centralised, involves a flood of request/response messages as shown in Figure 6(a). This naturally leads to a significant proportion of available bandwidth being used for management data.

On the other hand, recently reported NM MAFs [7][8][[9] assume a flat network structure. Hence as the number of managed devices grows, the network becomes increasingly unmanageable. This is a consequence of having a single MA responsible for obtaining NM data from each device in every PI (see Figure 6(b)), causing serious scalability problems. In addition, the order in which managed nodes are visited is arbitrary. This represents a significant problem when the management of remote LANs is considered, as a travelling MA may have to be transferred several times across expensive and low-bandwidth WAN links during its lifetime.

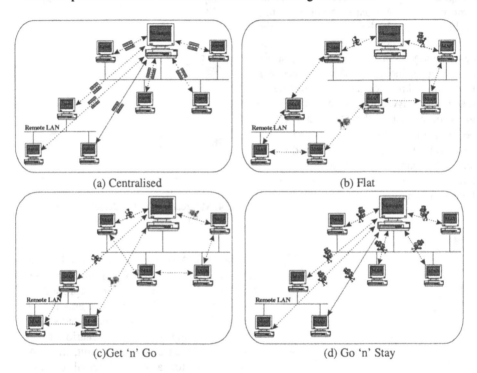

(a) Centralised

(b) Flat

(c)Get 'n' Go

(d) Go 'n' Stay

Fig. 6. Alternative approaches to polling.

As previously mentioned, the concept behind GnG polling is to partition the managed network into several logical/physical domains. The partitioning criteria are specified by the administrator and may correspond to: (i) the number of nodes assigned to each MA, (ii) the physical distribution of polled devices, or (iii) a hybrid of these two approaches. The number of MAs required per PI is automatically evaluated and their individual itineraries either manually or automatically specified. For instance, in Figure 6(c), an MA object polls the devices of the remote LAN, whereas a second MA is assigned to the network segment local to the manager host. With the GnG approach, the launched MA objects travel and perform their management tasks in parallel, whilst the number of devices they visit is limited to help minimise the overall response time. This factor suggests GnG as a suitable polling scheme for the acquisition of real-time data.

As the number of devices assigned per MA is reduced, the volume of MAs required to conduct polling is increased and the journey time of each of them decreased. Nevertheless, the manager needs more time and consumes more CPU cycles to instantiate, launch and receive back this larger number of MAs, whilst the communication overhead imposed by polling increases. Thus, an optimal solution has to be found as discussed in Section 6.

In terms of implementation, GnG polling is carried out through *Polling Threads* (PT). PTs are started and controlled by the manager application; each of them corresponds to a single polling instance. When started, PTs retrieve polling definitions and schedules from their associated configuration files and poll NEs on a regular basis. Specifically, PTs instantiate and launch the required number of MAs (supplied with their corresponding itinerary) and then sleep for one PI. When this period elapses the same process is repeated. Meanwhile, a manager's listener daemon receives the MAs that return to the manager carrying their collected data.

PTs may be synchronised such that they are not initiated simultaneously. This ensures that the traffic around the manager host will be distributed over time. In addition, the discovery of a new active agent process triggers an automatic re-evaluation of the required number of MAs per PI and their itineraries.

A graphical table (within the manager's UI) displays and allows modifications to existing PT properties, such as PT activation/de-activation, polling frequency and the number of devices assigned to each MA.

3.2. Go 'n' Stay Polling Mode

GnS polling introduces an alternative approach to NPM, targeting data intended for off-line analysis. This reduces the number of MA transfers whilst the volume of data carried by each MA increases. Hence, the proportion of *useful* management information within the MAs state is substantially increased, compared to the other approaches where a large number of MA transfers may be required to obtain a few data samples.

Specifically, every PT broadcasts at regular intervals an MA object to all agent hosts. The MAs then remain active on the hosts for p PIs (where p is specified by the administrator). At the end of each PI they obtain a sample of the requested data set and encapsulate it into their state. The MAs then sleep for one PI and awaken to obtain another sample. When the p PIs elapse, the MAs return to the manager to deliver the acquired samples (see Figure 6(d)). Meanwhile, PTs suspend execution for a duration given by the product of PI and the number p of PIs that MAs remain on the managed devices ($PI \times p$). When this period expires, they resume operation and the process is repeated.

Clearly, there is a trade-off between bandwidth consumption and response time. As p increases, so does the response time to the manager. However, as the MA transfers become sparser the network overhead imposed by polling reduces. If, for instance, $p=100$, MA objects will be broadcasted every 100 PIs. When changing to $p=50$, the MA transfers are doubled but the response time is halved. In the extreme case that $p=1$, the response time is minimised and GnS mode becomes similar to SNMP-based polling and identical to GnG, when each MA is assigned to a single device.

The administrator is given the option to modify the value of p and also dynamically change the polling mode from GnS to GnG and vice versa, depending on the managed network traffic conditions and the types of management information required. When changing from GnS to GnG, automatic network domain segregation takes place.

4. Semantic Compression

Polling is a frequent operation in NM as there are often several object values that require constant monitoring. Cases often occur however, where one or two Management Information Base (MIB) variables are not a representative indicator of system state and hence an aggregation of multiple variables is required (known as a *health function*) [16]. For instance, *five* MIB-II objects [3] are combined to define the percentage $E(t)$ of IP output datagrams discarded over the total number of datagrams sent during a specific time interval,

$$E(t) = \frac{(ipOutDiscards + ipOutNoRoutes + ipFragFails)*100}{ipOutRequests + ipForwDatagrams} \tag{1}$$

where MIB-II is an example of a MIB being supported by all the SNMP managed NEs.

In SNMP, the manager can only retrieve atomic MIB object values, hence all the operating agents will receive *five* requests corresponding to the *five* object values appearing in Eqn. (1). The value of $E(t)$ is then computed by the manager when all object values have been returned in separate response packets.

On the other hand, the MAs constructed by the MAG tool are able to perform *semantic compression* of management information. Thus, the value of $E(t)$ is computed locally, with a single value returned to the manager station. Hence, the manager is relieved from processing NM data, while the MAs state size remains as small as possible.

A major contribution made by this MA-based approach to NM is the *domain level* view of the managed network that mobility intrinsically implies. This is due to the multi-node movement that MAs often undertake, which allows for a superjacent level of data filtering. For instance, the manager requests the minimum value of $E(t)$ within the network: In such a case, each MA would compare the value of $E(t)$ acquired from the current host with the minimum value recorded so far and retain the lesser of the two.

Aside from the use of health functions, many other techniques could be applied to achieve data compression. For example, MAs could be customised to apply intelligent search methods to the extraction of data from large SNMP tables.

5. Performance Analysis

As evidenced in [11], a key factor affecting performance is the overhead induced by transport layer protocols. In particular, TCP is more reliable and yet traffic intensive resulting from its connection-oriented nature. In contrast, connectionless UDP sacrifices reliability in favour of a lightweight communication mechanism. In our implementation, the choice of the transport protocol is left to the network operator.

First, let us consider SNMP-based polling for the computation of Eqn. (1). If $S_{req/res}$ is the average request/response size (in application layer), and polling of n devices for v operational variables is applied, the wasted bandwidth for i PIs would be:

$$B_{SNMP} = 2 \cdot (S_{req/res} + O_T) \cdot n \cdot v \cdot I \qquad (2)$$

where O_T is the overhead imposed by the transport protocol (UDP in the SNMP case).

When GnG or GnS polling is employed, a considerable compression in data volume is achieved at the source, as explained in the preceding section. Thus, assuming an MA with compressed bytecode of size C and compressed state information of size S_h (when migrating from the hth host), the resulted overhead for GnG polling when segmenting the network into d domains would be,

$$B_{GnG} = n * (C + O_T) + d * i * \sum_{h=1}^{h_{tot}} (S_h + O_T), \text{ where } d \leq n \text{ and } h_{tot} = \left\lceil \frac{n}{d} \right\rceil + 1 \qquad (3)$$

as the v variables are aggregated into one. The first term of the equation describes the overhead imposed when broadcasting the MA code to all MASs, whilst the second represents the bandwidth consumed by the MA state transfers between the manager and the polled devices (each MA is assigned to $\left\lceil \frac{n}{d} \right\rceil$ NEs, hence h_{tot} transitions in total, including the return to the manager). Thus, for large v and i, GnG mode is less bandwidth intensive than SNMP-based polling. However, there is of course a threshold value of the number v of requested operational variables, below which the efficiency of the MA approach is lost in favour of SNMP-based polling [10].

The MA's state size, S_h, is given by,

$$S_h = S_{per} + S_{d_h} + S_{it_h} \qquad (4)$$

where S_{per} is the static/permanent information, while S_{d_h} and S_{it_h} are the data and itinerary vector's size respectively, when leaving the hth host. As explained in Section 2.4, S_{d_h} increases and S_{it_h} decreases for larger values of h. By defining the *constant* difference, dS, in MA's size after visiting each NE: $dS = (S_{d_h} + S_{it_h}) - (S_{d_{h-1}} + S_{it_{h-1}})$, Eqn. (4) becomes,

$$S_h = S_{per} + S_{d_1} + S_{it_1} + h * dS, \qquad \text{or} \qquad S_h = S_1 + h * dS \qquad (5)$$

and Eqn. (3) changes to:

$$B_{GnG} = n*(C+O_T) + d*i*\left[(S_1+O_T)*h_{tot} + \sum_{h=1}^{h_{tot}} h*dS\right], d \le n,$$

$$h_{tot} = \left\lceil \frac{n}{d} \right\rceil + 1$$

(6)

Similarly to GnG, the network overhead for GnS polling would be,

$$B_{GnS} = n*(C+O_T) + n*\left[2*(S_1+O_T) + dS'\right]*\left\lceil \frac{i}{p} \right\rceil$$

(7)

as MAs remain on the managed devices for p PIs collecting an equal number of samples resulting in a state size increment of dS'. For a large p, the number of MA transfers is minimised and GnS mode becomes the most lightweight polling approach in terms of bandwidth consumption.

As far as the transport protocol overhead O_T is concerned, when TCP protocol is in use, a single MA transfer requires a TCP connection to be set-up and subsequently released, i.e. 6 TCP messages, including acknowledgements. Each TCP message is prepended by a 20 byte TCP header and encapsulated into an IP datagram that introduces a 20 byte header[3], namely: O_T = 240 bytes. For UDP, a single packet is used to transfer the MA's state. The UDP header is 8 bytes long and encapsulated into an IP datagram, hence: O_T = 28 bytes.

6. Experimental Results

The experimental testbed comprises a network of several Solaris and WinNT machines. We assume the physical distribution of managed devices in the network as arbitrary to these experiments.

6.1. Response Time

In Figure 7, GnG polling mode is compared to an SNMP implementation [17] in terms of the response time for the acquisition of the health function result of Eqn. (1). Figure 7(a) shows that the flat approach (using one MA) does not scale well as the number of NEs increases. This makes it necessary to partition the managed network into several domains in order to maintain lower response times over the SNMP-based polling scheme.

---

[3] The measurements have been performed over an Ethernet LAN, which introduces an additional 26 bytes overhead on each frame. However, the overall size is slightly larger as there is a minimum frame size requirement of 72 bytes.

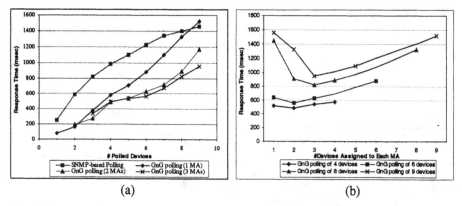

(a) (b)

Fig. 7. Response time of (a) SNMP-based vs. GnG polling, (b) GnG polling as a function of the number of devices assigned to each MA object.

Depending on the managed network size, the optimum number of domains can be determined from the minimum point of the corresponding curve of Figure 7(b). From the discussion in Section 3.1, when assigning a small number of devices to each MA, i.e. when launching a large number of MAs per PI, the overall response time increases. Specifically, although the individual journey times decrease, the time required for the manager to instantiate, launch and receive back these MAs dominates. Thus, for a network of six devices, the response time is minimised when each MA is assigned to two NEs, i.e. the network is segmented into three domains. Hence, the manager application (through the PTs) should autonomously adapt the number of domains according to the current number of managed devices. This issue is currently under investigation.

6.2. Bandwidth Consumption

Figures 8(a) and 8(b) illustrate a traffic overhead comparison between SNMP-based and GnG or GnS polling schemes respectively, according to equations (2), (6) and (7). For GnG and GnS polling, results for both TCP and UDP are shown.

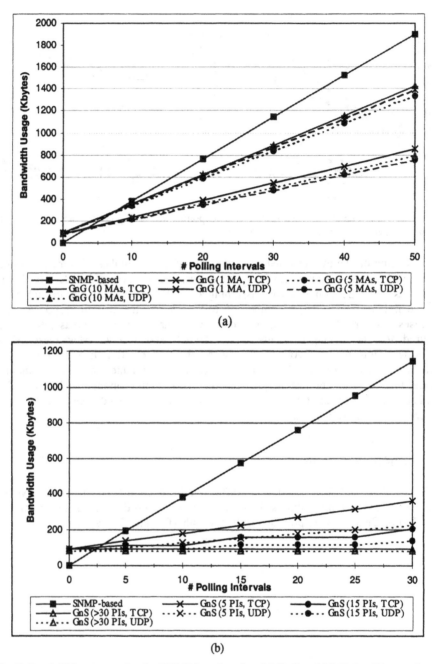

Fig. 8. Bandwidth consumption for SNMP-based against (a) GnG and (b) GnS polling modes.

For this experiment, $S_{req/res}$=50 bytes, C=1.6Kb, S_I=205 bytes, while n=50 devices and v=5 operational values. Also, the differences in state sizes were measured to be $dS = 3$ bytes and $dS' \approx 5$ bytes/sample. Hence, the initiation point for GnG/GnS polling depends on the transport protocol used and is derived from the first term of Eqn. (6) and (7). Clearly, both GnG and GnS modes represent a notable improvement over SNMP-based polling. However, the transport protocol choice has a very serious effect on bandwidth usage, with UDP significantly outperforming TCP.

An interesting result from the GnG polling experiment is that using a single MA does not necessarily prove less traffic intensive than using a larger number of MAs (see Figure 8(a)), although in the former case the overall number of MA transitions is minimised. This observation is explained by the summation: $\sum_{h=1}^{h_{tot}} h * dS$, included in Eqn. (6). In the flat approach ($d=1$), the MA object visits n hosts, with its state rapidly growing, causing scalability problems (for a large n). In contrast, when partitioning the network into several domains, each of the MAs return to the manager to deliver the results before its state becomes too large. Again, there is an optimum number of MAs that minimises the bandwidth consumption, dependent on the network size.

Figure 8(b) confirms that GnS mode becomes more attractive as the number PIs (p) that MAs remain on devices is increased. Hence, the choice of the transport protocol is crucial only for a small p, i.e. when MA transfers are relatively frequent. As the value of p increases the associated curves for TCP and UDP are shown to converge.

It can therefore be concluded that the selection of the appropriate polling mode (and the associated parameters, including the transport protocol), is a compromise between network overhead, response time and reliability, depending primarily on the type of management data to be collected.

7. Conclusions

A framework that exploits the capabilities of MAs in NPM and improves on the scalability of recently reported MAFs, has been presented. Security and fault-tolerance features have been integrated into the framework, and the efficiency requirement has been answered by introducing two novel polling modes.

First, we introduced the GnG mode that may be used for obtaining real-time NM data as overall response time is minimised. This method may also be preferable when considering the management of remote LANs. Concerning the off-line analysis of management data, we propose the GnS mode. In this approach, MAs collect a larger amount of data before returning to the manager leading to a direct reduction in the number of MA transfers.

In both cases, techniques for semantic compression of NM data are applied to reduce the bandwidth usage. The performance analysis and results indicate a significant improvement in both response time and traffic overhead when comparing the introduced polling modes to traditional centralised polling. The choice of transport protocol used for MA transfers has proven a critical factor regarding the polling modes' performance.

Current work addresses:

- Extensions to MAG functionality, such as constructing MAs able to filter SNMP tables;
- Optimisation of MAs itinerary to minimise GnG polling response time;
- Minimisation of MA state.

References

[1] Case J., Fedor M., Schoffstall M., Davin J., "A Simple Network Management Protocol (SNMP)", RFC 1157, May 1990.

[2] ISO/IEC 9596, Information Technology, Open Systems Interconnection, Common Management Information Protocol (CMIP) – Part 1: Specification, Geneva, Switzerland, 1991.

[3] McCloghrie K., Rose M., "Management Information Base for Network Management of TCP/IP-based internets: MIB-II", RFC 1213, March 1991.

[4] S. Waldbusser, "Remote Network Monitoring Management Information Base", RFC 1757, February 1995.

[5] Yemini Y., Goldszmidt G., Yemini S., "Network Management by Delegation", Proceedings of the 2nd International Symposium on Integrated Network Management (ISINM'91), 1991.

[6] Baldi M., Gai S., Picco G.P., "Exploiting Code Mobility in Decentralised and Flexible Network Management", Proceedings of the 1st International Workshop on Mobile Agents (MA'97), pp. 13-26, 1997.

[7] Ku H., Luderer G., Subbiah B., "An Intelligent Mobile Agent Framework for Distributed Network Management", Proceedings of the IEEE Global Telecommunications Conference (Globecom '97), pp. 160-164, 1997.

[8] Susilo G., Bieszczad A., Pagurek B., "Infrastructure for Advanced Network Management based on Mobile Code", Proceedings of the IEEE/IFIP Network Operations and Management Symposium (NOMS'98), pp. 322-333, 1998.

[9] Gavalas D., Greenwood D., Ghanbari M., O'Mahony M., "An Infrastructure for Distributed and Dynamic Network Management based on Mobile Agent Technology", accepted to IEEE International Conference on Communications (ICC'99), 1999.

[10] Gavalas D., Greenwood D., Ghanbari M., O'Mahony M., "A Hybrid Centralised - Distributed Network Management Architecture", accepted to the 4th IEEE Symposium on Computers and Communications (ISCC'99), 1999.

[11] Fuggetta A., Picco G.P., Vigna G., "Understanding Code Mobility", IEEE Transactions on Software Engineering, 1998, vol. 24, no. 5, pp. 342-361.

[12] Sun Microsystems: "Java Language Overview – White Paper" [On-line] (1998), URL: http://www.javasoft.com/docs/white/index.html.

[13] Sun Microsystems, Jini Technology White Papers (1999), URL: http://sun.com/jini/whitepapers/.

[14] Sun Microsystems, Java Object Serialisation Specification, Feb. 1997.

[15] Rivest R.L., Shamir A., Adleman L., "A Method for obtaining Digital Signatures and Public-Key Cryptosystems", Communication of the ACM, 21(2), Feb. 1978.

[16] Goldszmidt G., "On Distributed Systems Management", Proceedings of the 3rd IBM/CAS Conference, 1993, URL: http://www.cs.columbia.edu/~german/papers.html.

[17] AdventNet, URL: http://www.adventnet.com/.

Realization of an Agent-Based Certificate Authority and Key Distribution Center

Karsten Bsufka, Stefan Holst, and Torge Schmidt

DAI-Lab, Technical University Berlin, Germany
{kbsufka|stefamig|schmidt}@cs.tu-berlin.de
http://dai.cs.tu-berlin.de/

June 17th, 1999

Abstract Security issues are key factors for the deployment and acceptance of agent based systems in the telecommunication area. This fact is most obvious in electronic commerce applications, where security services have to be offered. These services are needed to ensure secure communication, fair exchange of goods and payment. Public key cryptography techniques are an often employed mechanism. Keys are distributed by using a certificate to store them and to provably associate them to a principal. This document deals with the design of an agent-based certificate authority (CA) and key distribution center (KDC).

1 Introduction

At the DAI laboratory of the Technical University Berlin, we are developing the Java-based agent architecture for telecommunication applications JIAC (Java Intelligent Agent Componentware [1]).[1]

Secured services and security services are a key factor for the acceptance of telecommunication applications, especially in the area of electronic commerce. One important part of the security architecture for JIAC are certificates[12], they are used on all levels of the architecture. Because certificates are widely used, it is necessary to provide mechanisms for the creation, administration, and distribution of certificates. Existing mechanisms for working with certificates were not efficient or useful enough for the use in JIAC.

A variety of work has been done and is going on to standardize secure communication between agents[5, 21]. Although these attempts use public key cryptography, they don't deal with the problem of realizing an agent-based public key infrastructure (PKI). Instead, they simply assume there is access to a PKI, which can be used by agents. Also, they don't describe a standardized way to communicate with certificate authorities (CA) or key distribution centers (KDC).

In this document we want to show how we realized an agent-based CA and an agent-based KDC. Furthermore, we show why this realization is independent

---

[1] The project is funded by the Deutsche Telekom Berkom GmbH.

from any existing PKI or CA concept. It can be used as a generic approach to different application scenarios and as a base infrastructure.

This text is arranged as follows: after a short general introduction to JIAC and it's security infrastructure follows the design description of the agent-based KDC and after that the design of the CA. The last section gives an outlook into the future.

2 JIAC Overview

JIAC is a Java-based agent architecture for realization of telecommunication applications[1]. A JIAC agent consists of individual components implemented as JavaBeans[3], with a fixed set of exchangeable core components and an optional set of application dependent components.

JIAC agents use KQML based speech acts to communicate with each other[4], although it is planned to migrate to FIPA-ACL based communication shortly. For the remainder of this document we stick with the KQML design and terminology.

The abilities of an agent are described by plans. Plans describe preconditions for the execution of a service and define a postcondition (effect), which will be fulfilled, if a plan was successfully executed. The service name and the service preconditions will be used as a description of the service, available externally.

In JIAC, each agent runs in an agent execution environment. These environments are grouped into an agent community, called marketplace. Several marketplaces form one organisational domain.

3 Security Infrastructure for JIAC

The security infrastructure for JIAC offers security mechanisms for agents and agent environments on different levels, which is depicted in figure 1:

- Java Virtual Machine
- Transport Layer (ISO/OSI Layer 4)
- Agent
- Agent communities

Certificates are used on all of these levels. It is necessary to define an ontology for security, this ontology can then be used by JIAC agents. This ontology has to include all aspects of security offered by JIAC, and each agent which wants to use security mechanisms has to understand this ontology. In order to enable conformity with agent standards, JIAC offers the possibility of adding additional ontologies to an agent, these additional ontologies should enable the agent to communicate securely with a (e.g.) FIPA agent.

The techniques used in the security infrastructure are all based on established algorithms and protocols[19], and only standard cryptography packages will be used, like the Java Cryptography Architecture and the Java Cryptography Extension (a non- American re-implementation)[13, 14, 11].

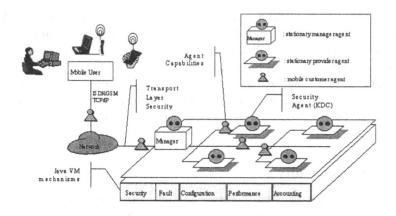

Figure1. Security in JIAC

3.1 Java Virtual Machine

In the first stage of development of the security infrastructure for JIAC, the mechanisms of the Java 2 security architecture will be used for securing the Java Virtual Machine (VM)[10, 18, 9].

For adding agents to an agent environment, a class loader for agents has been implemented, it is based on the URLClassLoader of Java 2 [8, 17]. Class loaders will be used for the migration of mobile agents and when adding stationary agents to an agent environment.

The security architecture of Java 2 uses certificates for the assignment of permissions to applications/applets. They form a part of the code source, which is used for the definition of a protection domain within the VM[7]. The certificates are stored in "keystores", administration is done by a command line tool. For defining a security policy, the Java security API uses certificates from the keystores. However, Java 2 contains no mechanisms for the distribution and verification of certificates. Therefore the JIAC architecture will provide graphical certificate administration and distribution tools.

3.2 Transport Layer

JIAC offers the possibility to add any available communication driver to an agent, each communication driver is treated as an additional component.

In order to secure communication on the transport layer, JIAC offers a SSL communication driver, which can be added to each agent[6], thereby implementing session-oriented security.

3.3 Agent

Each agent can have additional cryptographic capabilities: If a communication component for security on the transport-layer is unavailable or not wanted, it is also possible to secure the communication on the agent communication level. Speech acts may either be secured separately for asynchronous communication, or within a protocol session. The required abilities and certificates will be provided as an optional core component.

As another functionality, on the service level a security component checks each service request the agent receives for its autorization to use the requested service. This is done by consulting a service control list for a matching autorization entry.

3.4 Agent community

JIAC uses certificates to identify agents and agent environments, and to check if a mobile agent has the permission to migrate to another agent environment or to accept the migration of an agent from another agent environment.

3.5 Agent-based CA and KDC

Since certificates will be used on the agent and on the agent community level, it is necessary to provide agents, which are capable of distributing and verifing certificates. In the following, we will introduce the agents in the security infrastucture of JIAC, which are responsible for certificate management. One will implement a KDC and the other a CA.

Existing methods for certificate administration A set of standards for a PKI exists which are all based, more or less, on X.509[12, 15, 2]. Standardized formats for certificates are available and are in use with real world applications, but a widespread, available, and useful PKI for the verification and distribution of certificates is still missing.

Usually, existing CAs (e.g. VeriSign, ...) offer the possibility to request or distribute certificates over the WWW. However, agents can't use these procedures easily in an automated way. Further, CAs normally use a proprietary protocol for the verification and distribution of certificates (e.g. TTP from TeleSec[20]). Standardized certificate revocation mechanisms are lacking, too.

Java 2 offers rudimentary ways for the administration of certificates, but no procedure for the distribution and verification of certificates is integrated.

Design goals A designated agent, called security agent, is responsible for security issues in the JIAC architecture. Scalability of the design allows the security agent to be used hierarchically on each architectual level: on the agent execution environment, the agent community, or the organizational domain level.

A security agent functions also as a mediator between a CA and agents, by acting as a KDC. Services are defined within the JIAC security ontology, that allow agents to deliver certificates to a KDC or to verify certificates.

This mediator role can be accomplished in any of two ways, depending on the scaling decision of the installed system. Either, access components for specific

CAs are added to the security agent, or dedicated wrapper agents are contacted by it. Though, the service which is offered to client agents, is the same for both approaches. One component will be provided, that accesses the agent-based CA as implemented by the project.

Security Agent (KDC) The security agent acts in the role of a KDC. Besides providing KDC functionalities, a security agent offers trust management functionalities to the agent community infrastructure as well as to mobile agents, but that is beyond the scope of this document. The security agent offers the following services as a KDC:

- Certificates look up, with a variable list of search parameters.
- Verification of certificates.
- Request of all valid certificates.
- Providing an up to date CRL.
- Forwarding to a CA:
  - Revocation of certificates.
  - Certification Requests conforming to PKCS #10[16].

The security agent is always called KDC for the remainder of the document, since only this functionality is discussed here.

Agent-based CA Agent The agent-based CA agent fulfills the following functions:

- Support of X509v3 certificates.
- Reading certificates from a file or Java 2 keystore.
- Storing certificates into a file or Java 2 keystore.
- Creation of new certificates.
- Processing of PKCS #10 Certification Requests.
- Adding/removing certificates into/from the CA agent knowledge base.
- Distribution of certificates to KDCs.
- Validation of certificates.
- Management and distribution of CRLs.
- Providing a graphical user interface, which can be used by a CA administrator.

The following two sections describe the agend-based KDC Agent and the CA Agent in more detail way.

4 Key Distribution Center (KDC)

In this section we give an overview of all functionalites a KDC agent offers to other agents, and which speech acts can be used to request a service. The interface to the CA agent is described in section 5.

4.1 Administration of certificates

This section contains the overview of all KDC agent services related to certificates.

Retrieving an existing certificate Other agents should have the possibility of inquiring existing certificates from a KDC.

To search for existing certificates, the following speech act can be used:

```
(ask-one :content (ask-certificate <params>))
```

The parameter **params** contains a list of pairs, each of these pairs consists of the identifier for a certificate attribute and a value for this attribute. The receiving KDC searches for matching certificates. If several pairs of attribute and value are given in the request, these are linked with a logical and. Certificate attributes which are not part of the request are interpreted as wild cards, and they match all entries.

A KDC are using the following speech act to report back the search results:

```
(reply :content (certificates-found
                 <certificates>))
```

In the answer the parameter **certificates** contains all certificates, which matched the search request. No revoked certificates are returned.

Deleting a certificate It is also possible for other agents to request the deletion of a certificate. Before deleting the certificate, the KDC has to check if the sender of the request has the permission to delete a certificate, which is implemented by using the service control list mechanism as described in section 3.3. The request to delete a certificate is sent by using the following speech act:

```
(evaluate :content (remove-certificate
                    <certificate>))
```

In the speech act the parameter **certificate** contains the certificate, which should be deleted by the KDC.

4.2 Certificate Revocation Lists

This section contains the overview of all KDC agent services which deal with certificate revocation lists.

Requesting a CRL With the following speech act the most current available CRL will be requested from a KDC:

```
(evaluate :content (request-crl))
```

The CRL will be sent with the following speech act:

```
(tell :content (crl <crl>))
```

The parameter **crl** contains the most current CRL known to a KDC.

Testing a certificate for revocation To simplify communication and to reduce overhead an agent can ask the KDC to return the status of a certifcate.

Therefore it utilizes the following speech act:

```
(ask-one :content (test-certificate <certificate-id>))
```

The KDC replies with the following speech act:

```
(reply :content (certificate-status <status>))
```

status can be either unknown for a non-existing certificate, if it fails validity checks it is invalid, revoked if it was revoked, or valid otherwise. Note that it is sensible to use authentication upon performing this message exchange.

4.3 Communication between a KDC agent and a CA

Each KDC agent has special components to communicate with real world CAs like VeriSign or TeleSec, or it has the ontologies and protocols necessary to communicate with an agent-based CA as described in section 5. A KDC has only components and ontologies for that CAs, which are known to him.

5 Agent-based Certificate Authority Agent

Now follows an overview of CA agent functionalities and a description how these services can be requested by using speech acts. For the regular case we assume the CA to be reachable on-line, which might lead to possible security threats. This approach has been chosen for demonstration purposes of the prototype.

5.1 Administration of certificates

We will now describe the agent-based CA agent, or CA agent for short. The CA agent administers all certificates, which are known by him. The CA agent differentiates between valid, invalid, and withdrawn certificates. A CA agent must be able to produce new certificates, to process certificate inquiries, and to distribute certificates. Likewise, it must be possible to withdraw certificates and request the current CRL.

Creating a certificate The data needed for creating a certificate should arrive at a CA agent in a secure way. The transport of data can take place off-line or on-line. If the data is transmitted to the CA agent off-line, the certificate is then produced by using the user interface of the CA agent, usually after verifying the data.

The certification request which is sent to a CA agent, must conform to the PKCS #10 (Certification Request Syntax) standard.

In order to request a certificate online, a secure speech act must be transmitted to the CA agent, i.e. the speech act must be signed and encrypted. This means, the sender must have generated a public-private key pair already, and the sender must know the CA certificate. In the case of restrictive software licensing a valid product certificate might be required also. The distribution of the CA certificate, and optionally the product certificate, is not handled by the security infrastructure of JIAC. The CA certificate could be part of a distribution and the setup application will generate the key pair upon installation. To request the issue of a certificate the following speech act will be used:

```
(ask-one :content (request-new-certificate <request>))
```

For the syntax of the parameter **request** the standard PKCS #10 is used. The certificate will be sent back to the requestor, who is contained within the originating speech act. The whole content of the request must be signed by the sender with its newly generated private key and must be encrypted with the public key of the CA agent.

If a certificate was issued, it will be transmitted using the following speech act:

```
(reply :content (new-certificate <certificate>))
```

certificate contains the issued certificate. The response is signed by the CA agent.

Distributing certificates The graphical interface of the CA agent can be used to enter a list of KDCs, to which a list of all available valid certificates should be sent automatically.

To send only a new certificate to a KDC, the following speech act is used:

```
(evaluate :content (add-certificate <certificate>))
```

The following speech act is sent by the CA agent to all known KDCs, if the CA agent wants to transmit all available valid certificates:

```
(tell :content (certificate-list <certificates>))
```

5.2   Certificate Revocation Lists

The CA agent offers the possibility to revoke a certificate and to request the most current certificate revocation list (CRL). The services to revoke a certificate and to request a CRL are also offered by a KDC to a third agent, which forwards the request appropriately.

Revoking a certificate An agent/user must have the possibility to revoke a certificate he owns. To do this, he sends the following speech act to the CA:

```
(evaluate :content (revoke-certificate <certificate>))
```

In the speech act the parameter `certificate` contains the certificate, which should be revoked. This speech act has to be signed by the owner of the certificate, and the service control mechanisms of the CA must permit the action.

If a certificate was successfully revoked, the following speech act will be sent:

```
(tell :content (revoked-certificate))
```

Distributing a CRL To all KDCs which are known by the CA agent a CRL will be sent, if a certificate is revoked or a fixed time interval has passed. To sent a CRL the following speech act is used:

```
(tell :content (crl <crl>))
```

The parameter `crl` contains the most current CRL.

5.3   CA Agent user interface

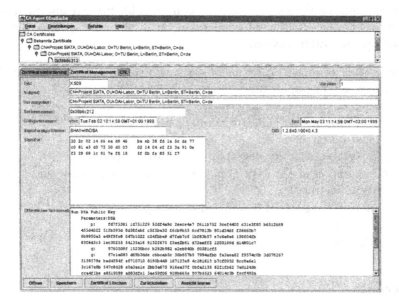

Figure2. GUI of CA agent

The CA agent has a graphical user interface (GUI) to configure and control the CA agent (see figure 2). The GUI can be used to create, view, revoke, import, export, and delete certificates. It is also possible to import certification requests conforming to PKCS #10 [16] and to create certificates using these requests. Certificate revocation requests might be read in from a file, too.

Lists of all certificates and CRLs can be sent manually to all known KDCs. Likewise, it is possible to configure, when certificates or CRLs should be sent automatically to all KDCs. The addresses for the KDCs can also be entered by using the GUI.

6 Conclusions and Future Work

In this document was presented how a generic CA agent will be realized, that can encapsulate the functionality of any existing CA. Likewise, it was shown how an agent-based KDC can be realized.

The uniform way to communicate with an agent-based CA agent or KDC enables an agent to use certificates without the need to possess the knowledge about the specific realization of a CA and the protocols used by this CA for requesting and validating certificates. It is a practical example for distributed problem solving by communication and delegation to autonomous entities.

Beyond the use in prototype systems it is necessary to achieve a co–operation with a real world CA, so that the respective provider is able to implement a CA agent, or dedicated CA agents can be supplied for them.

Future work on the security infrastructure will be directed towards the improvement and the extension of existing functionalities, among other things the following points will be covered:

- Integration of smart card readers, for smart cards that store private keys and have cryptographic functions.
- Provision of value added services:
  - Arbitrator and adjudicator agents/components.
  - Time stamp services.
  - Proxy agents for cryptography and security services.
- Implementation of the Java PolicyProvider class, especially tuned for our security infrastructure:
  - Implementing a GUI to add and remove permissions during run-time of an agent environment.
  - Achieving independence from policy files.

References

1. Sahin Albayrak and Dirk Wieczorek. Jiac — an open and scalable agent architecture for telecommunication applications. In H. Velthuijsen and S. Albayrak, editors, *Intelligent Agents in Telecommunications Applications — Basics, Tools, Languages and Applications.* IOS Press, Van Diemenstraat 94, 1013 CN Amsterdam, The Netherlands, January 1997. ISBN 90 5199 295 5.

2. Alfred Arsenault and Sean Turner. Internet X.509 Public Key Infrastructure PKIX Roadmap. Located at http://www.ietf.org/internet-drafts/draft-ietf-pkix-roadmap-00.txt, September 1998. PKIX Working Group. Internet Draft.

3. Graham Hamilton (Editor). JavaBeans™ Version 1.01, July 1997.

4. Tim Finin and Jay Weber et al. Draft specification of the KQML agent comminication language. Located at http://www.cs.umbc.edu/kqml/kqmlspec/spec.html, June 15 1993.

5. FIPA. FIPA 98 Specification Part No.10, Version 1.0 Agent Security Management. Located at http://www.fipa.org, October 1998. FIPA - Foundation for Intelligent Physical Agents.

6. Alan O. Freier, Philip Karlton, and Pail C. Kocher. The SSL Protocol Version 3.0. Located at http://home.netscape.com/eng/ssl3/index.html, March 1996.

7. Li Gong. Implementing Protection Domains in the Java Development Kit™ 1.2. In *Proceedings of the Internet Society Symposium on Network and Distributed System Security*, pages 125–134, San Diego, California, March 1998.

8. Li Gong. Secure Java Class Loading. *IEEE Internet Computing*, 2(6):56–61, November/December 1998.

9. Li Gong, Marianne Mueller, Hemma Prafullchandra, and Roland Schemers. Going Beyond the Sandbox: An Overview of the New Security Architecture in the Java Development Kit™ 1.2. In *Proceedings of the USENIX Symposium on Internet Technologies and Systems*, pages 102–112, Monterey, California, December 1997.

10. James Gosling, Bill Joy, and Guy Steele. The Java Language Specification. Located at http://java.sun.com/docs/books/jls/html/index.html, 1996. Sun Microsystems, Inc.

11. IAIK-JCE. Located at http://jcewww.iaik.tu-graz.ac.at/. IAIK Java Security, Institute for Applied Information Processing and Communications, Graz University of Technology.

12. ITU-T. Recommendation X.509: The Directory–Authentication Framework, June 1997.

13. Java™ Cryptography Architecture API Specification & Reference. Located at http://java.sun.com/products/jdk/1.2/docs/guide/security/CryptoSpec.html, October 1998. Sun Microsystems, Inc.

14. Java™ Cryptography Extension 1.2 – API Specification & Reference. Located at http://java.sun.com/products/jce/. Sun Microsystems, Inc.

15. Steve Kent. RFC 1422 Privacy Enhancement for Internet Electronic Mail: Part II: Certificate-based Key Management, February 1993.

16. RSA Laboratories. PKCS #10: Certificate Request Syntax Standard. Located at http://www.rsa.com/rsalabs/pubs/PKCS/, November 1993.

17. Sheng Liang and Gilad Bracha. Dynamic Class Loading in the Java Virtual Machine. To appear in the 13th Annual ACM SIGPLAN Conference on Object-Oriented Programming Systems, Languages, and Applications OOPSLA'98, October 1998. ACM.

18. Tim Lindholm and Frank Yellin. The Java™ Virtual Machine Specification. Located at http://java.sun.com/docs/books/vmspec/html/VMSpecTOC.doc.html, September 1996. Sun Microsystems, Inc.

19. Bruce Schneier. *Applied Cryptography*. Joh Wiley & Sons, Inc, 2nd edition, 1996.

20. TeleSec. Analyse einer TTP-Message. Located at http://www.telesec.de, Dezember 1998. Deutsche Telekom AG.

21. Chelliah Thirunavukkarasu, Tim Finin, and James Mayfield. Secret Agents — A Security Architecture for the KQML Agent Communication Language. Draft submitted to the CIKM'95 Intelligent Information Agents Workshop, October 1995.

Providing Telecommunication Services through Multi-agent Negotiation

Mihai Barbuceanu[1], Tom Gray[2] and Serge Mankowski[2]

[1]Enterprise Integration Laboratory
University of Toronto
4 Taddle Creek Road, Rosebrugh Building,
Toronto, Ontario, Canada, M5S 3G9
mihai@ie.utoronto.ca
[2]Mitel Corporation
350 Legget Drive, P.O. Box 13089
Kanata, Ontario, Canada K2K 1X3
{tom_gray,serge_mankowski}@mitel.com

Abstract. To provide services to customers, the components of the international telecommunications system have to agree on joint activities and on the use of resources. In this paper we show how simple and clear models of interaction and behavior can be combined in a generic negotiation architecture able to automate the agreement reaching process between agents that, like in the global telecommunications system, have different preferences, policies and authority over resources and activities. At the basis of the architecture is an authority model of how agents can influence each other by setting obligations and interdictions upon their behavior. This ability is conferred to agents by their ownership over resources and activities. The architecture integrates a conversational component, enforcing and ensuring the well-formedness of the interaction, a representation of action formalizing agents' authority to set obligations and interdictions upon other agents and a constraint optimization reasoning component allowing parties to deliberate over behaviors and outcomes to decide on their next move. We discuss in detail how this architecture is applied to dynamically negotiate the provisioning of communication services based on the detection and resolution of feature interactions.

1 Introduction

The international telecommunications system is the largest distributed computing system in the world, composed of many autonomous subsystems with different ownership over resources and activities. To provide services to customers, these subsystems have to agree on joint activities and on the use of resources. But reaching such agreements is often hampered by unexpected and undesired interactions among the service features preferred by the various parties. These delay the introduction of new features to the market, complicate development and maintenance and cause dissatisfaction among users. In this paper we describe a generic negotiation architecture able to automate the agreement reaching process between the various actors who use or provide communication services, given their specific preferences, policies and ownership over resources and activities.

At the basis of the architecture is a model of *authority* stipulating how agents can *influence* each other. We believe that an agent's power to influence other agents stems from the agent's ownership over resources and activities that are needed by other agents to achieve their goals. Agents with ownership over such resources and activities can charge and reward other agents for the use of their resources and activities (like e.g. telephone companies charging for the use of the call display service, but also giving rewards like free use of call display during the first month). To represent this in a general manner we take a decision theoretic approach assuming that in general agents are utility maximizers, that is they will generally adopt those behaviors that will maximize their expected utility (some deviations from this will be discussed later on).

We represent an agent's power to influence other agents by using utility based definitions of *obligations* and *interdictions*. Agents set obligations and interdictions for other agents by charging them (decreasing their utility) if they adopt undesired behaviors or by rewarding them (increasing their utility) if they adopt desired behaviors. For example, a customer (with ownership over his business) may forbid the service provider to ever connect him to certain 1 900 numbers. If the provider fails to achieve this, the customer's dissatisfaction may determine him to scale down his business with the provider, leading to a net loss of utility to the provider. Or a company may have an obligation towards VIP customers to handle their calls within a guaranteed reduced response time. Obligations and interdictions may come in various degrees of strength. Strong obligations and interdictions may be very costly to violate (like the interdiction to be connected to certain 1 900 numbers), while weaker ones may be violated at a lesser loss (like when even VIP customers are put on hold for a longer time if the system load is very heavy).

The architecture, implemented as a generic *negotiation shell*, integrates representation, reasoning and coordination models and components as follows.

1. A *behavior representation* component, allowing agents to represent their own goals and behaviors, as well as model the others' goals and behaviors. Obligations and interdictions are represented relative to the *roles* that agents play in the interaction process, and are quantified in terms of costs and rewards associated with the execution or non-execution of behaviors.

2. A *reasoning and decision making* component, allowing agents to deliberate about goals, behaviors and outcomes. To know what to propose, accept or reject agents need to determine, understand and quantify the consequences of given actions and behaviors on both their own and others goals. We support this by means of a *constraint optimization engine* that searches in behavior spaces dynamically generated based on previous knowledge and on knowledge obtained through interactions. This provides *guarantees* for the optimality of the behaviors adopted by agents.

3. A *process* component, that deals with the structure of the interactions taking place among parties, ensuring the well-formedness of the dialogue. Parties *request, propose, accept reject*, etc. goals and behaviors according to a process following shared conventions, until some agreement is reached or some failure condition holds. We support this by means of *conversational mechanisms* that provide the descriptive and execution framework for communicative action based structured interactions.

The paper starts with presenting the components of the architecture, namely behavior representation, the search engine and the conversational interaction support. Then we show how this can be applied to negotiate service provisioning in telecommunications by dynamically detecting and solving feature interactions. We end with related work and concluding remarks.

2 Describing Behavior

Syntax and Semantics. Agents act and thus we first need a language to describe and reason about agent behavior. Our language contains two types of actions, *composed* and *atomic*. Composed actions consist of other (sub)actions, while atomic actions do not. Both types of actions have [earliest-start, latest-end] execution time intervals within which they must be executed (discrete time assumed). Atomic actions have specified finite durations. We allow three kinds of compositions.

- *Sequential* compositions, $a = seq(a_1, a_2, ...a_n)$ denote that all component actions a_i must be executed in the given order, inside the time interval of a (their superaction), without temporal overlapping.

- *Parallel* compositions, $a = par(a_1, a_2, ...a_m)$ denote that all component actions must execute within the time interval of a, but with temporal overlapping allowed.

- *Choice* compositions, $a = choice(a_1, a_2, ...a_p)$ denote execution of only a nonempty subset of sub-actions within the time window of the super-action, also with overlapping allowed. *Exclusive choices* (*xchoice*) also require the execution of at most one component action.

From the execution viewpoint, choices have *or* (*xor* for xchoices) semantics in that a choice g is 'on' - meaning will be executed and written $On(g)$ - iff at least one component is on (only one for xchoices) and 'off' - meaning will not be executed and written $Off(g)$ - iff all components are off. Sequences and parallels both have *and* semantics - 'on' iff all components are on, and 'off' otherwise. The difference is that sequences also require *ordered* execution of subgoals, while parallels don't. We address this in a manner that is logically equivalent to breaking any sequence with more than two elements into several two element (binary) sequences and by introducing specific inference rules for the binary sequences. For example, the sequence $s = seq(a_1, a_2, a_3)$ is treated as if it were defined as $s = par(seq(a_1, a_2), seq(a_2, a_3))$. Then, if $On(s)$ that means we must have $On(a_1, a_2)$ *and* $On(a_2, a_3)$. ($On(a, b)$ obviously means a is executed before b). As a specific inference rule, $On(a_i, a_{i+1}) \supset Off(a_{i+1}, a_i)$. From this we also derive that if a sequence is 'on', all its subsequences are also 'on', and if a subsequence is 'off' then all its super-sequences are also 'off'. E.g., $Off(a_1, a_3) \supset Off(a_1, a_2, a_3)$.

Obligation and Interdiction. Setting obligations and interdictions is the way by which agents use power upon each other. There are two ways of obliging or forbidding an action. An obligation can be created by imposing a cost if the action is not done or by giving a reward if it is done. Similarly, an interdiction can be created by imposing a cost if the action is done or giving a reward if it is not done. We use the notation

$O_-(g,c)$ to express that g is obliged with cost c, $O_+(g,r)$ for g being obliged with reward r and similarly for interdictions, $F_-(g,c)$ and $F_+(g,r)$. In terms of the on-off state, if an action is on, then all interdiction costs are paid and all obligation rewards are accumulated, while if it's off all obligation costs are paid and all interdiction rewards are accumulated.

Non-power utilities. Setting obligations and interdictions is not the only way to influence the behavior of another agent. When the relations among agents do not allow the use of authority arguments, requesting agents may simply quantify the importance (from their viewpoint) of some goals, leaving it to the other agent to decide if and to what extent the request can be satisfied. For example, $u(g,v)$ quantifies the utility of $On(g)$ as v (with g negated it would be the utility of $Off(g)$) from the viewpoint of some requesting agent, without explicitly setting any obligation or interdiction w.r.t. g.

Roles. Finally, agents also have a representation of roles that specifies, for each role an agent can play or interact with, the goals that can be obliged or forbidden and possibly the range of costs and rewards that the role has authority to use. This is needed to check the legality of using power arguments in interaction. In our system each agent has such a representation and performs these checks when appropriate. Alternatively, these verifications could be delegated to an external trusted agent, especially if this is also going to be used as an arbiter, mediating disputes about rights and authorities.

3 Searching for Optimal Behaviors

Terminology. Addressing now the issue of how the above representations can be used to search for optimal agent behaviors, let's first define some terminology.

Let $G = \{g_1, ...g_n\}$ be a goal (action) network. An on-off labeling of the network is mapping $L : G \rightarrow \{on, off\}$ associating either 'on' or 'off' labels to each action in G. A labeling is *consistent* iff the labels of each composed node and of its subgoals are consistent with the node's execution semantics, e.g. if a parallel goal is 'on', then all its subgoals are also 'on' or if a choice goal is 'off', then all its subgoals are 'off', etc. Consistent labelings thus define *executable* behaviors.

Let $C = \{c_1, ...c_m\}$ be a set of constraints, where each c_i is a constraint of the form $On(g_j)$, $Off(g_k)$ or an implication on both sides of which there are conjunctions of on-off constraints. For example, $On(OnSiteService) \supset On(PayOnSiteService)$ is an implication.

Let $P = \{p_1, ...p_k\}$ be a set of obligations and interdictions. Each p_i is one of $O_-(g_i, c_i)$, $O_+(g_i, r_i)$, $F_-(g_i, c_i)$ or $F_+(g_i, r_i)$. This is the set of 'power arguments' currently considered. Given a set P and a consistent labeling L of G, we can compute $Cost_{L,P}(G)$ as the sum of all interdiction costs for 'on' goals and all obligation costs for 'off' goals and respectively $Reward_{L,P}(G)$ as the sum of all interdiction rewards for 'off' goals and all obligation rewards for 'on' goals. (We will drop subscripts where clear). Having both rewards and costs gives agents more flexibility in expressing how they choose to influence one another. For the purpose of the search for optimal behaviors (see next), rewards and costs are translated into utilities, just like the non-power utilities, and used in the formulation of the optimization criterion.

Finally, $U = \{u_1(g_1, v_1)...u_l(g_l, v_l)\}$ is a set of (non-power) utilities, communicated by another agent. Given a set U and a consistent labeling L of G we can compute $Util_{L,U}(G)$ as the sum of all (non-power) utilities for the labeled goals in G.

Searching for Behavior. Our main method of reasoning about behavior uses a search method to find a consistent labeling (or behavior) L over G that satisfies all constraints in C and optimizes or satisfies a given *criterion* formulated in terms of the elements of P and U: $Search(G, C, P, U, criterion) = L$.

The criterion is formulated by the agent in response to a request from another agent or to its own needs and reflects the social attitude that the agent adopts vis-a-vis the request or its own situation. Normally, a *social* agent would maximize its own $Reward(G) - Cost(G)$ while seeing to it (to the extent possible) that others' $Util(G)$ are still above some threshold. More cooperative agents may at times want to maximize others' $Util(G)$ while keeping their $Reward(G) - Cost(G)$ above some threshold. In the latter case, if the cost incurred for maximizing the requester's utility is too high, the agent may demand a compensation from the requester, in the form of another action that the requester should achieve. This leads to the creation and exchange of arguments of the type 'I'll do X for you if you'll do Y for me'. To find an appropriate 'Y', an agent can look at its outstanding goals over which the other agent has control. Otherwise, 'Y' can always be 'Pay me something', with the amount to be paid being dynamically computed from the difference between $Cost(G)$ and $Reward(G)$.

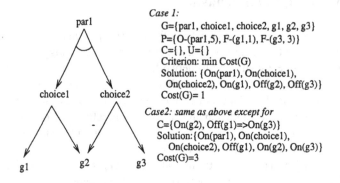

Case 1:
G={par1, choice1, choice2, g1, g2, g3}
P={O-(par1,5), F-(g1,1), F-(g3, 3)}
C={}, U={}
Criterion: min Cost(G)
Solution: {On(par1), On(choice1),
   On(choice2), On(g1), Off(g2), Off(g3)}
Cost(G)= 1

Case2: same as above except for
C={On(g2), Off(g1)=>On(g3)}
Solution:{On(par1), On(choice1),
   On(choice2), Off(g1), On(g2), On(g3)}
Cost(G)=3

Fig. 1. Behavior Search

We can also represent 'deviant' attitudes, like: (1) *Devoted*: maximize $Util(G)$ regardless of $Cost(G)$ and $Reward(G)$, (2) *Selfish*: maximize $Reward(G) - Cost(G)$ regardless of $Util(G)$, (3) *Anti-social*: minimize $Util(G)$ or (4) *Self-destructive*: minimize $Reward(G) - Cost(G)$ regardless of $Util(G)$.

The search method is illustrated in figure 1 (choice1 and choice2 are choices, par1 is a parallel, the rest are atomic, and g2 occurs negated in choice2). In the first case, without constraints and rewards and minimizing $Cost(G)$, the optimal behavior has cost 1, incurred by $On(g_1)$ that violates $F_-(g_1, 1)$. In the second case, with constraints, the min cost behavior has cost 3, due to $On(g_3)$ which violates $F_-(g_3, 3)$.

Algorithms. In the general framework described, consistent on-off labeling of arbitrary goal networks with constraints is intractable (being equivalent to satisfiability) and so is the associated optimization problem. The main search algorithm we use is a complete *branch-and-bound* backtracking search method. This guarantees that the best behavior will be found, and has acceptable performance for moderate size networks.

4 Managing the Process

To support the process dimension of negotiation we use our previous conversational technology [Barbuceanu &Fox 97]. As this is described elsewhere, we only review here a few elements needed for the understanding of this work. The major elements of our conversational technology are *conversation plans, conversation rules, actual conversations* and *situation rules*. Briefly, a conversation plan is a description of both how an agent *acts locally* and *interacts* with other agents by means of communicative actions. The specification of conversation plans is largely independent from the particular language used for communication for which we currently use a liberal form of KQML [Finin et al. 92]. A conversation plan consists of states (with distinguished initial and final states) and rule governed transitions together with a control mechanism and a local data base that maintains the state of the conversation. The execution state of a conversation plan is maintained in *actual conversations*. To decide which conversations to instantiate and to update the agent's data base when events take place, we provide *situation rules*. The top level control loop of an agent activates all applicable situation rules (suggesting conversations to initiate) and then executes new or existing conversations as appropriate.

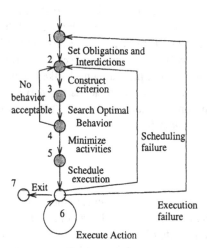

Fig. 2. Basic plan searching and executing behaviors

A top level behavior for an agent may simply consist of the agent setting its relevant obligations and interdictions, deciding on a search criterion and applying it to search for the optimal behavior. If this behavior contains goals that require cooperation with other agents, new sub-conversations are spawned to negotiate the other agent's commitment. The initial conversation is suspended until the sub-conversations terminate, after which is resumed. The conversation plan in figure 2 shows one possible incarnation of this scheme. It shows how:

1. An agent receives utilities representing obligations, interdictions and other preferences that another agent requests it to satisfy (state 1).

2. Translates, verifies and installs them in its own goal network (state 2).

3. Decides on its attitude vis-a-vis the request and constructs the search criterion (state 3).

4. Runs the search engine (state 4).

5. Minimizes the actions to be executed in the optimal behavior. This is done by computing subsumption relations $(On(a) \supset On(b))$ among actions and then applying a min-cover algorithm. The purpose is to avoid executing actions that are implicitly achieved by other actions. (state 5).

6. Schedule the required actions (state 6). Scheduling determines the final order of actions that satisfies all sequencing and resource constraints (we allow these on actions too) as well as the time windows in which each action has to be executed, and is provided by means of a constraint based scheduler. Scheduling failures determine a rerun of the search method with a different criterion.

7. Execute the scheduled actions (by rule execute-action in state 6). If during action execution new constraints are posted, propagations will be redone. The action execution service is carried out by an executive that makes sure the real time is within the time window for each action and monitors the success or failure of each action.

8. If no execution failure occurs, the plan ends in state execution-ok, otherwise it ends in execution-failed. If during behavior search the agent determines that it can not satisfy some of the requested constraints, or if actions can not be scheduled or planned, the violated constraints may be sent back to the sender for revision.

This also illustrates our basic approach of providing tools like behavior search, action translation, minimization, scheduling, execution etc. as *services* that may or may not be used by an agent, depending on the situation it is in. These services are made available to the agent through a *Behavior API*.

5 Negotiation Scenarios

Most of the value delivered by telecommunication systems to their users comes from allowing users to tailor the telecommunication system's behavior according to their needs [Cameron et al. 96]. These include aspects like which users to include in calls, where to route calls, what to do when parties are unavailable, which resources to use, etc. But these needs often conflict. A user may not want to be connected to certain numbers, but when she calls another subscriber her call may be forwarded to one of the undesired numbers. Or a user may not want to accept calls from certain numbers, but if she was unavailable when a call from one of these numbers arrived, the automatic recall service may later call back the unwanted number. The telecommunication industry currently solves these interactions in a centralized manner inside multi-million lines of proprietary software. This has severe problems. Service providers must determine how combinations of features will behave, including combinations with other providers features. Maintaining and extending the feature management software when new features are being added is very hard, especially since the correct support of features often depends on the intentions behind their use by various parties. A better solution is thus to have decentralized agents programmed by users handle the interactions by run time negotiation [Griffeth & Velthuijsen 93]. This would enable users to specify their individual policies involving calls, without being aware of others policies.

To evaluate the power of our negotiation technology, let us now show how it can be used to build such a solution. We start with a few basic examples of the features that are usually available (see [Cameron et al. 96] for a classification of features):

- *Incoming Call Screening*: the callee will refuse all calls from callers in an incoming call screening list.

- *Call Forward*: the callee will forward the incoming call to another number.

- *Retry*: if the callee is busy, either the the caller or the callee will try to establish the connection later.

- *Outgoing Call Screening*: the caller does not allow to be connected to some specified directory numbers.

As discussed, the feature interaction problem arises from combinations of features that interact in undesired ways, affecting or obstructing the intended functionality of the provided services. *Incoming Call Screening* and *Retry* may conflict if *Retry* is done without checking that the number belongs to the incoming call screening list - we shouldn't call back numbers that are not accepted in the first place. Similarly, *Call Forward* and *Outgoing Call Screening* may conflict if a caller is forwarded to a number that it does not wish to be connected to.

Assume now A and B are agents responsible for establishing voice connections amongst their users. Figures 3, 4 and 5 show goal structures used by agent B when receiving a call from A. Note that the boxed goals in the figure are goals that can only be achieved in collaboration with agents playing the roles attached to the boxes. In other words, the agent does not have complete discretion over the execution of these goals and has to negotiate with others to carry them out.

Scenario 1: Basic feature interaction. Consider figure 3 and assume A has *Outgoing Call Screening* for numbers #1 and #2, and B has *Incoming Call Screening* with A in its incoming call screening list (meaning B does not want to talk to A directly). The set of constraints that A sends to B is

```
{ (O_ Take-my-call 5)
  (F_ ch1 9)
  (define choice ch1 Forward-to-Moldova Forward-to-Hong-Kong) }.
```

When receiving this message, B first translates, verifies and installs the requested behavior. Goal Take-my-call is translated to ReceiveCall, while Forward-to-Moldova

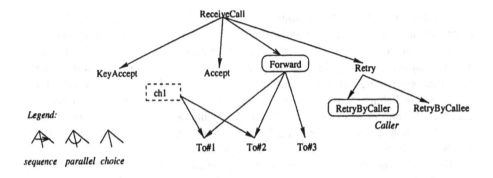

Fig. 3. Basic call processing behaviors.

and `Forward-to-Hong-Kong` are translated to `To#1` and `To#2` respectively and a new goal, `ch1` is installed among B's goals (in interrupted lines in figure 3). As A is on B's screening list, B will also post it's own interdictions for `Accept` and for `Retry`. The search engine is run and finds a behavior that does not violate any of the stated obligations and interdictions, namely $\{On(To\#1), Off(To\#2), Off(To\#3),$ $Off(Accept), Off(KeyAccept), Off(Retry)\}$. A is thus forwarded to `To#1`, because the other two possible forwarding numbers are forbidden by A and `Accept`, `KeyAccept` and `Retry` are forbidden by B. This behavior is immediately executable as it does not violate any constraint.

Alternatives to this scenario include:

1. A has `Outgoing Call Screening` to all of #1, #2 and #3. In this case it is not possible to satisfy all requests, but if A has different violation costs for the three numbers, B can forward to the smallest cost one. This may be done automatically by B or with A's approval, in which case the negotiation cycle is extended to include the approval step.

2. A does not want B to know that it does not want to be forwarded to #1 and #2. In this case, A does not send the corresponding interdiction, but requires B to confirm any forwarding. B replies by informing A about all three possible numbers to forward to and A either calls one of these directly (so that B doesn't know which it is) or asks to be forwarded to a specific one (so that it is not clear which number was avoided).

Scenario 2: Protecting against unwanted calls. In figure 4 assume that B has a *Key Protection* feature that requires that certain callers input a key before being connected to B. If A is among these callers, B will place an obligation for `KeyAccept`. Assuming also that A is not one of the callers whose call is accepted only in the presence of a lawyer (see next) and that B is not away, interdictions will be also placed on `IncludeLawyer` and `Reroute`. Again, this has a zero violation behavior consisting of A entering the key and, if successful, being connected to B.

A more sophisticated form of protection is allowing calls from some callers only if the callee's lawyer is also included in the call. This is an example of a *custom* service

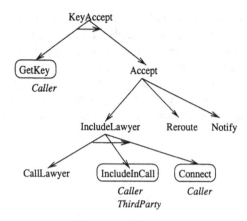

Fig. 4. Protecting against unwanted calls.

that some subscribers may create themselves. Assume that A is among those callers whose call is only taken in the presence of B's lawyer. Detecting this forces B to place an obligation on `IncludeLawyer`. The zero violation behavior in this case has B try first call his lawyer and, if successful, take A's call. If the lawyer is not available, her agent may schedule a future time for the call. This time will be communicated to B who in turn will counter propose it to A. If agreed, A will know when to call next time to be able to talk to B.

We note that forbidding or obliging various goals depends on checking a variety of conditions about the agents' state, including agents' own databases that contain screened out callers, callers required to enter valid keys, or even callers for whom a lawyer's presence is required. We provide several ways in which these verifications can take place, including situation rules, conversation rules or specific rules attached on goals that determine if the current situation warrants posting obligations or interdictions on the goal.

Scenario 3: Security concerns. Figure 5 deals with security concerns when transferring calls to the user's current location (the `Reroute` goal). When a user is at a location considered insecure either by the user or by a caller, the call processing agent must make sure that the call does not take place. As neither the callers' insecure locations nor how much risk they are willing to accept can be known in advance, they have to be communicated dynamically. This happens by allowing callers to include in their initial request definitions for the `TransferCallerSafe` goal as, e.g., parallel compositions where the disallowed goals appear negated. The strongest way to ensure that the called party will not ignore these security requirements is to set costs for the interdictions associated with the insecure locations, under the assumption of power relations allowing callers to charge called parties if their security requirements are ignored:

```
{(parallel TransferCallerSafe
  (- TransfLocation1) (- TransfLocation2))
 (F_ TransfLocation1 9)
 (F_ TransfLocation2 5)}
```

The called party's agent will set to 'on' the `TransfLocationj` goal corresponding to the actual location of the user and search for a *minimum* cost `AcceptCall` behavior. If this cost is greater than zero because of violating caller interdictions, it will be communicated to the caller. If the caller is willing to take the risk expressed by the cost, it will explicitly require so, acknowledging that the called party is not responsible for this cost anymore. Otherwise, only a notification will be sent to the user. This is a clear example of how the dialogue is driven by the exercise of the power that parties have been endowed with.

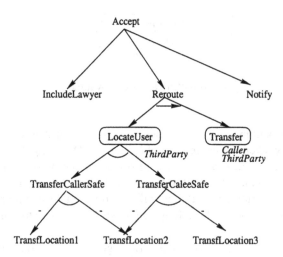

Fig. 5. Transferring calls, with security concerns.

6   Current Status, Related and Future Work

We have completed the Java implementation of the presented model, consisting of (1) classes for representing goals, constraints and utilities, (2) an optimized branch-and-bound complete search method, (3) methods for checking the consistency of sequences, (4) a simple Pert-like scheduler allowing durations and start/end times, to be replaced in the near future by a full fledged constraint based finite capacity scheduler and (5) a real time execution environment that executes goals according to the schedule. This has been integrated with a new Java implementation of our conversation coordination language [Barbuceanu & Wai-Kao 99]. This combination provides a complete Java platform for implementing complex negotiation scenarios of the type illustrated in figure 2. The resulting system is being integrated into an agent oriented telecommunications platform by Mitel Corp.

With respect to related work, [Parsons, Sierra & Jennings 98], have recently proposed a logical approach to negotiation by having agents reason internally by formal argumentation. Technically, this relies on free theorem proving in multiple modal logics. This is general in principle, but the practical difficulty of direct implementation by theorem proving, which they imply, is not dealt with. More importantly, we differ from them with respect to what we consider to be the basis of argumentation. Their approach is to have an agent accept an argument of another party if the agent can not logically contradict it. For example, they illustrate a case where an agent A gives a tool T1 to an agent B when B successfully argues that A can also do its job with another tool T2. We believe that in reality even if B can use T2, this does not necessarily mean that he has to accept the request to give T1 *unless* there is some *utility* to be gained (or lost) if he does it (or respectively if he doesn't). In other words, we believe that at the foundation of argumentation lies an evaluation of the *utility* the agent stands to gain or loose and not logical refutability *per se*. In particular, we can easily conceive negotiations where agents may accept arguments even if logically they may appear incorrect, if coming from agents exercising enough power - like when an employee acts to satisfy the CEO's request, even if he does not agree with its logic.

One can also argue that reasoning and building arguments about this utility can itself be encoded in the logical language(s) of the agent and thus make the refutability approach applicable in this case as well. This is true but leaves the construction of the actual mechanism for reasoning about utility entirely on the shoulders of agent designers. We have tried to remove this burden by providing such a general mechanism. This makes agents easier to construct and provides *guarantees* about the optimality of their behavior.

On the feature interaction front, [Griffeth & Velthuijsen 93] present a negotiation approach that shares many intuitions with ours. Where we differ is in the higher level of formality and inference power offered by our negotiation solver and in the richer infrastructure that we offer. Griffeth and Velthuijsen provide the notions of abstraction and composition for complex action specification. These appear similar to our choice and sequence types, but their meaning seems to come more from the underlying Prolog implementation than from a separate formal specification. Similarly, the inference power of their method is not clear and neither is the complexity of reasoning (although they imply their reasoning is tractable).

We have previously done work on using deontic logic in coordination [Barbuceanu 98], proposing an arc-consistency propagation method to infer the consequences of given obligations and interdictions. The price for tractability was a less expressive action language, the apparent impossibility to consider constraints and very limited querying capabilities. This experience has shown that deontic logic is too limited in the range of behaviors it can describe. Thus, we have taken the opposite route, removing these limitations at the price of higher complexity. We believe that realistic negotiation can not do without these extended capabilities.

A first limitation that we are working on comes from the computational cost of systematic search as done by our search engine, which may become problematic as we increase the size of goal networks. Given the equivalence of on-off labeling to satisfiability, we are building a local search engine based on satisfiability methods [Selman, Levesque & Mitchell 92], with heuristics biasing the solution toward improving the given criterion. This can be used either stand alone, or to provide a good initial upper bound for the complete *branch-and-bound* method.

Second, we know that for goals that require another's agent participation, the commitment of the other agent must be negotiated first, and this negotiation may fail. There is a danger that agents may spend too many resources trying to negotiate behaviors that seem advantageous to them but don't have enough chances to succeed (like trying to negotiate a long distance call free of charge). The cause of this is the failure to consider the *probability* of achieving such goals. We are addressing this by using Bayes networks [Pearl 88] to represent agents' causal models of actions and to compute the probabilities of actions that agents don't control completely from observations and assumptions about the actions that agents do control. These probabilities are used to calculate more informed *expected* costs, rewards and utilities which are then used by the search engine to determine the best course of action.

7 Acknowledgements

This research is supported, in part, by Materials and Manufacturing Ontario, Mitel Corp., Communications and Information Technology Ontario, Natural Science and Engineering Research Council, Digital Equipment Corp., Micro Electronics and Computer Research Corp., Spar Aerospace, Carnegie Group and Quintus Corp.

References

[Barbuceanu &Fox 97] Barbuceanu, M. and Fox, M. S. 1997. Integrating Communicative Action, Conversations and Decision Theory to Coordinate Agents. *Proceedings of Automomous Agents'97*, 47-58, Marina Del Rey, February 1997.

[Barbuceanu 98] Barbuceanu, M. 1998. Agents that work in harmony by knowing and fulfilling their obligations. *Proc. of AAAI-98*, Madison, WI, 1998.

[Barbuceanu & Wai-Kao 99] Barbuceanu, M. and Wai-Kao, L. 1999. Conversation Oriented Programming in COOL. M. Greaves and J. Bradshaw (eds.) *Proceedings of Autonomous Agents'99 Workshop on Specifying and Implementing Conversation Policies*, Bellevue WA, May 1999.

[Cameron et al. 96] Cameron, E.J., N.D. Griffeth, Y.J. Lin, M.E. Nilson, W.K. Schnure, and H. Velthuijsen. 1996. A Feature Interaction Benchmark for for IN and Beyond. In L.G. Bouma and H. Velthuijsen, editors, *Feature Interactions in Telecommunication Systems*, 1-23, Amsterdam, IOS Press.

[Finin et al. 92] Finin, T. et al. 1992. Specification of the KQML Agent Communication Language. The DARPA Knowledge Sharing Initiative, External Interfaces Working Group.

[Griffeth & Velthuijsen 93] Griffeth, N. and Velthuijsen, H. 1993. Win/win negotiation among autonomous agents. *Proc. of 12th Workshop on Distributed Artificial Intelligence*, Hidden Valley, PA, 187-202.

[Pearl 88] Pearl, J. 1988. Probabilistic Reasoning in Intelligent Systems, Morgan Kaufmann.

[Parsons, Sierra & Jennings 98] Parsons, S., Sierra, C., and Jennings, N. 1998. Agents that Reason and Negotiate by Arguing. *Journal of Logic and Computation*, 8(3), 261-292.

[Selman, Levesque & Mitchell 92] Selman, B., H.J. Levesque and D. Mitchell. 1992. A new method for solving hard satisfiability problems. Proc. AAAI-92 San Jose, CA, pp. 440446.

Dynamic VPN Provisioning through Communicative Agents

Danny Jacxsens and Bart Bauwens

Corporate Research Center, Alcatel
F. Wellesplein 1, B-2018 Antwerp, BELGIUM
{Danny.Jacxsens, Bart.Bauwens}@alcatel.be

Abstract. Due to the liberalisation of the telecom market, the capability of controlling complicated business processes has become more important than ever for deploying telecom services. Also, inter-operability between heterogeneous systems and networks has become a key issue. Furthermore, the need to cope with a variety of terminals, or to offer personalised services in a rapid way, are all becoming essential components of a successful business plan. Intelligent agents bring us an attractive way to solve many of those problems. Standardisation of Agent Communication Languages (ACLs) is however a first prerequisite of success. One of the most relevant standard bodies in this area is the Foundation for Intelligent and Physical Agents (FIPA). The European FACTS project has as primary objective to validate and contribute to FIPA via a number of application trials, including one in the domain of Virtual Private Network (VPN) provisioning. This paper will first describe the different types of agents and the VPN provisioning scenarios. The FIPA ACL specification has been used to implement those scenarios, but standardising only on an ACL is not sufficient. One should also agree on vocabularies used to describe services and user profiles. This paper will show the advantages of using content languages such as XML and RDF for service transactions and descriptions.

1 Introduction

In today's landscape of open telecom markets, telecom operators have started to play new roles. At the same time completely new players are appearing on the market. The business interactions between those roles have become much more complicated than before. The liberalisation process has also resulted in a mixture of heterogeneous systems, networks and terminals. On the other hand, telecom companies still want to meet the high expectations of their customers, asking for on-demand, robust and personalised services.

Software agents offer an interesting approach to solve many of the above issues. A lot of discussion material already exists on what exactly is meant with 'software agents' [1]. For the sake of this discussion, we can simply think of agents as being autonomous pieces of software, which behave in the interest of their users. In this paper, we will only focus on intelligent and communicative agents, and not discuss the use of mobile agent technology.

2 VPN Provisioning through Communicative Agents

Communicative or Intelligent Agents can take over the burden of the complex interaction mechanisms between different players, such as negotiations or charging. Agents can easily represent one of the business roles such as backbone operator, access provider or end-user, and act on their behalf, based on established policies or user profiles. Intelligent Agent technology is also well suited to customise services during the service lifetime. By exchanging messages, the interfaces between end-user and service provider can be dynamically changed to optimise the interactions.

In the context of the European FACTS project, we applied the agent paradigm to the provisioning process of Virtual Private Networks (VPN). FACTS is validating and contributing to several agent-related standard bodies such as FIPA, OMG and W3C, based on the experiences gained in case studies in three different areas: audio-visual entertainment and broadcasting, network provisioning and travel assistance. This paper will only focus on the network provisioning work.

As a first step, the roles of customer, VPN service provider and network provider were mapped to different types of software agents. The following assumptions about the future telecom environment were made:

- There will be typically multiple competing network providers (ISPs or telephone operators) each responsible for their own network domain, which allow subscribing to their connectivity services, possibly during a relatively short period of time.
- Service providers will take care of negotiating with and choosing between the different network providers. Their role could be taken by new players, but also by ISPs or operators themselves.

The most important agent types considered in this dynamic environment are listed below:

- A Personal Communication Agent (PCA) will represent the interests of its end-user as good as possible. The end-user will run one or more applications on top of the VPN which can be controlled each by a different Application Agent (see further on).
- VPN Service Provider Agent (VPN SPA) is a special type of Service Provider able to act as a broker between PCA and NPAs. It will look for the best network provider and offer a number of interesting choices to the PCA.
- A Network Provider Agent (NPA) represents a Network Provider responsible of managing its own network domain. Federation between different NPAs is also possible.

In the Section 3, we will go into more detail on the functionality offered by those agents. A few additional agents will be considered, which are useful to integrate with the user applications.

3 Dynamic VPN Provisioning Scenario

The VPN Provisioning scenario proposed by the FACTS project comprises multiple consecutive scenario stages, deemed necessary for co-ordinating and provisioning a multi-user application across a multimedia VPN. We will only discuss here the most important steps in this scenario, i.e. the Meeting Scheduling and Service Provisioning scenario stages. The complete scenario also covers aspects such as start-up, termination, reconfiguration and charging of the VPN service.

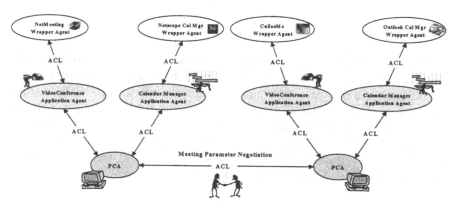

Fig. 1. Meeting Scheduling

In the Meeting Scheduling scenario stage (Fig. 1), the Personal Communication Agents act as personal organisers for the users, arranging a mutually convenient time for the multimedia application to be started. In this negotiation there is also interaction between the PCA and the Application Agents. These Application agents manage all applications or services of a specific type which are accessible to the user (in our case a Video Conferencing tool and a Calendar Management tool), and interact with the Application Wrapper Agents which represent the actual application (each application requires its own Application Wrapper Agent). Note that all agents in Fig. 1 interact via FIPA ACL messages (see Section 4).

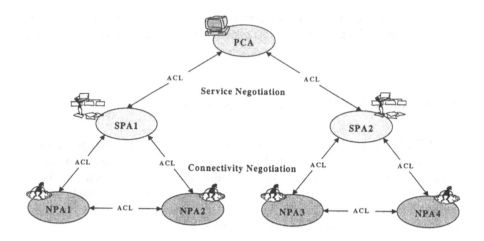

Fig. 2. Service Provisioning

When the Meeting Scheduling stage is finished, the Service Provisioning stage (Fig. 2.) starts. One of the PCAs will communicate the details of the required service to a selection of SPAs. Each SPA must then negotiate with a set of NPAs to arrange the provision of the required VPN connection. This results in the selection of an NPA, which meets best the requirements of the SPA. Following this the SPAs which were contacted by the PCA will send their service offer to the PCA, which finally selects the SPA which offer matches the service requirements best. All negotiations are based on parameters such as time/date/duration, quality of service, etc.

4   A Standardised Approach for VPN Negotiation

4.1   The use of FIPA's Agent Communication Language

Key success factor of software agents is the development of standard communication languages. The relevant standard body for communicative agents is the Foundation for Intelligent and Physical Agents (FIPA) [2], an international association of companies developing standards for agent technologies. FIPA agent communication is based on speech-act theory [3]: messages consist of a header (the 'act'), followed by the subject of the act (referred to as the 'content').

In the context of the VPN Provisioning case study of the FACTS project, we evaluated a subset of FIPA ACL acts, such as 'inform', 'request', 'propose' or 'cfp'. Apart from their impact on the agent's mental attitude, the ACL is useful to guide a dialog-based design, especially when combined with use-cases. Indeed, all interactions described in the above scenarios can be considered as conversations, using specific protocols. One of FIPA's strengths is also the ability to choose between a set of standard protocols such as the 'FIPA-request' protocol or the 'FIPA-iterated-contract-net' protocol.

However, it should be clear that standardising on an Agent Communication Language alone is not sufficient to achieve interoperability: of equal importance is the standardisation of VPN ontologies and content languages. In the next section we will describe the approach chosen within the FACTS project.

4.2 Ontologies for VPN Provisioning

4.2.1 Ontology Requirements

During the development, the main emphasis was on developing as much as possible open standards, which allow for real interoperability between different systems. The only format, which is still popular after more than ten years, is ASCII. And even if it may not always be the most efficient one, it is the most human-readable and portable format. Standardisation on the higher-application levels will always remain difficult to achieve and therefore reuse of already established specifications is very important. A modular approach seems to be most appropriate. In this spirit of openness and reusability, both the Web and the paradigm of OO programming have seriously influenced the development.

For the VPN Provisioning scenarios described in Section 3, we made a distinction between the following vocabularies:

- Service transactions: including service requests, offers, common service properties.
- Service descriptions: including service specific properties.

In addition to those, we defined some general-purpose vocabularies for describing:

- Quantities: special types of quantities are bandwidth, duration, ... together with their corresponding units.
- Events: type, start, end, duration of events.

4.2.2 XML-based Ontologies

The eXtensible Markup Language (XML) is a W3C Recommendation [4], which allows representing and exchanging structured information on the Web. As it is a meta-language, interested communities or industry domains can develop new languages or vocabularies by agreeing upon the definition of a DTD (Document Type Definition). The syntax of XML instances is based on the use of tags and attributes, in a way similar to HTML.

XML will dramatically influence the Web developments of the next few years. Currently a variety of Web vocabularies are emerging on the Web in very different domains such as e-commerce, finance, software deployment, mathematics, chemistry, etc. One may expect that a set of DTDs in the telecommunication domain will become available as well, resulting in de-facto standards for exchanging service, user and terminal profiles or in 'intelligent' application protocols. We believe that in order to create really successful agent systems today, one can not deny the serious impact of these developments. Therefore we have chosen XML as the underlying mechanism for the communication between our agents. As agent content language, XML also gives the perfect answer to the requirements described above, such as the need for reusability and openness.

XML can be combined with XSL style-sheets to create human-readable representations of agents' messages in ordinary Web pages. The next releases of major browsers Microsoft Internet Explorer and Netscape will include native XML support. In addition two related specifications XLink & XPointer can be used to specify links between parts of the content. This may be useful to identify parts of the content and refer in subsequent messages to those parts without repeating them.

A wide variety of XML supporting tools already exist both in the public domain as in the commercial world. Examples of such tools include parsers, browsers, editors, translators, or database engines. The major browsers also provide standardised APIs (as the DOM [5]) to manipulate or query the XML content.

4.2.3 FIPA and XML-based Content Languages

FIPA does not mandate the usage of one particular content language, but instead allows applications choosing an 'appropriate' one. There is however one important requirement: the language should be able to express 'actions', 'statements' and 'objects'. Languages such as KIF [6] have built-in support for expressing those concepts, but are only popular in closed academic circles. When using XML as content language, the meaning of the elements is usually specified in the documentation of the DTD.

As XML has no built-in support to represent propositions, actions, etc., one solution may be to let the DTD designer specify how the different element types can be mapped into those concepts. This mapping can also be formalized using mechanisms such as architectural forms. Anyway, establishing such a mapping for an existing DTD is not straightforward (if possible anyway).

XML DTDs impose a rigid tree structure, while graph-based representations are typically more flexible and better suited for 'knowledge' representation. XML-based efforts to represent knowledge are OML [7] and the more powerful CKML [8]. In the Section 4.2.4, we will discuss the usage of RDF as an alternative graph-based solution.

4.2.4 RDF as Content Language

The Resource Description Framework (RDF) [9] defines a mechanism for describing (Web) resources (meta-data), to enable "automated" processing of these resources. It provides a model for representing these meta-data, and proposes XML as serialisation syntax for this model. Using RDF Schema [10] a meta-model of the RDF data model can be defined (also in XML syntax). RDF allows describing conceptual model and is -in this respect- better suited as FIPA content language.

The objects and propositions as referred to in the FIPA specifications can be easily mapped to resources or properties in the RDF model. However, RDF has no built-in notion of actions: therefore, we developed RDF schemas to allow the encoding of:

- Actions (including the actor and the arguments)

- Results of those actions
- State information on the action, such as action 'done', 'refused', 'ongoing', 'interrupted', etc.

The following message illustrates how an initial Call For Proposals ('CFP') can be issued by a Personal Communication Agent (PCA) to a VPN Service Provider Agent (SPA) in order to set up a VideoConference service over a VPN, using FIPA ACL with RDF as content language. The example assumes that RDF schemas are available for the ontologies fipa (FIPA-extenstions), str (Service Transaction), sd (Service Description) and event , as specified by the XML namespaces.

```
CFP
  :sender     pca_1@iiop://www.geocities.com/acc
  :receiver   spa_1@iiop://www.operator.com/acc
  :language   RDF
  :ontology   VCMeeting
  :content
  ( <rdf:RDF
      xmlns:rdf="http://www.w3.org/1999/02/22-
              rdf-syntax-ns#"
      xmlns:fipa="http://www.fipa.org/rdfExtensions#"
      xmlns:str="http://www.alcatel.be/ontology/
              ServiceTransaction#"
      xmlns:sd="http://www.alcatel.be/ontology/
              ServiceDescription#"
      xmlns:event="http://www.alcatel.be/ontology/
              event#">

    <str:proposeVCService rdf:ID="initialProposal">
      <fipa:actor>
        spa_1@iiop://www.operator.com/acc
      </fipa:actor>
      <fipa:argument rdf:resource="proposedService"/>
    </str:proposeVCService>

    <sd:VPNService rdf:ID="proposedService">
      <sd:events rdf:resource="events"/>
      <sd:participants rdf:resource="participants"/>
      <sd:description rdf:resource="vcdescription"/>
    </sd:VPNService>

    <sd:EventBag rdf:ID="events">
      <sd:event_li rdf:resource="event1"/>
    </sd:EventBag>

    <sd:EventInfo rdf:ID="event1">
      <event:startDateTime>
        19990324T150000000Z
      </event:startDateTime>
      <event:duration>
        +00000000T010000000
      </event:duration>
    </sd:EventInfo>
```

```
<sd:VCDescription rdf:ID="vcdescription">
  <sd:serviceType> av-conference </sd:serviceType>
  <sd:serviceName> NetMeeting </sd:serviceName>
  <sd:profile rdf:resource="vcprofile"/>
</sd:VCDescription>

<sd:VCProfileInfo rdf:ID="vcprofile">
  <sd:profileType> video </sd:profileType>
  <sd:framerate rdf:ID="rate"
                sd:unit="FramesPerSecond"
                rdf:value="50"/>
  <sd:viewFrameSize rdf:ID="size"
                    sd:unit="Pixel"
                    rdf:value="33600"/>
  <sd:imageComplexity> color </sd:imageComplexity>
  <sd:noiseLevel rdf:ID="noise"
                 sd:unit="dB"
                 rdf:value="5"/>
</sd:VCProfileInfo>

<sd:ParticipantBag rdf:ID="participants">
  <sd:participant_li rdf:resource="participant1">
  <sd:participant_li rdf:resource="participant2">
</sd:ParticipantBag>

<sd:ParticipantInfo rdf:ID="participant1">
  <sd:userName>user1</sd:userName>
  <sd:emailAddress>usr1@org1.com</sd:emailAddress>
  <ipAddress>111.222.333.444</ipAddress>
</sd:ParticipantInfo>

<sd:ParticipantInfo rdf:ID="participant2">
  <sd:userName>user2</sd:userName>
  <sd:emailAddress>usr2@org2.com</sd:emailAddress>
  <ipAddress>555.666.777.888</ipAddress>
</sd:ParticipantInfo>

</rdf:RDF>)
```

The example shows how the verbosity of messages can be reduced by predefining some elements by unique XML identifiers, which can then be referred to in replies to this message. This mechanism may be especially useful when the negotiation focuses only on one specific service parameter.

5 Dynamic VPN Provisioning Test Environment

5.1 FIPA Agent Platform

The FIPA Agent Platform (Fig. 3) developed by Alcatel within the context of the European FACTS project, was based on the MASIF-compliant Agent Platform Grasshopper [11]. Extensions were added to the Grasshopper platform in order to make it compliant to the FIPA specifications. Those extensions include a number of agents providing White Pages, Yellow Pages and message routing services, an ACL-parser and an RDF- parser (based on W3C's SiRPAC [12] and IBM's XML4J [13] parsers). The RDF-parser parses the RDF Descriptions given in the ACL message content field and the RDF Schema's which are referred in the Descriptions and which are loaded at runtime from a Web server. It generates a set of RDF-triples, which the RDF Datamodel software validates and transforms into an RDF Graph object model. This RDF Graph consists of a set of linked RDF Nodes (each of them a Java object) which represent the RDF resources form the Description and their interrelationships. This graph structure can be queried easily by the agent software.

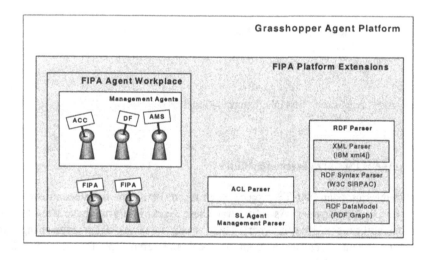

Fig. 3. FIPA Agent Platform

5.2 Agent Application Test Environment

The whole test environment of the application described in Section 3, is spread over three international sites interconnected via Internet and an ISDN Network (Fig. 4). At each of the sites a similar test environment is set up, consisting of a router/firewall combination, a set of PCs/Workstations on which the Agent Platform and the Agents reside, and a Web server on which the RDF Schema's can be placed.

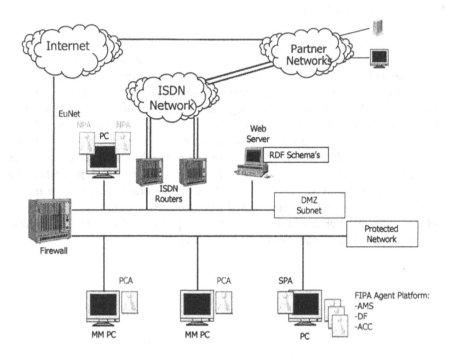

Fig. 4. Agent Application Test Environment (Alcatel Site)

5.3 Agent Platform Interoperability

Within the European FACTS project, three Agent Platforms have been developed by three different companies. As a first trial these Agent Platforms have been tested for interoperability over the internet. These trials involved agent registering, agent searching, agent deregistering both on local and remote platforms, and federated searching over all platforms.

6 Conclusions

Within the context of the European FACTS project, we aimed to demonstrate a number of VPN provisioning scenarios in an open telecommunication environment. Agents representing the different players communicate in the FIPA Agent Communication Language. To ensure real interoperability between different systems, services or applications standard content languages are needed. We argued that XML technology suits very well to agent technology and have shown how to use it for the defining a set of ontologies in the area of service provisioning, with an emphasis on VPNs. The use of the FIPA ACL justified the use of the RDF model, resulting in a set of RDF Schemas.

We believe these efforts are certainly a first step forward towards service negotiation and interoperability. However, still many questions remain unanswered so far, such as how does RDF compare with other schema initiatives, such as Xschema [14], the Document Content Description (DCD) [15], or what is the exact relation with UML. Finally, a framework for classification of telecom services would be needed, to allow the combination of similar efforts for defining telecom ontologies and achieve inter-working between different services.

7 References

1. "Software Agents: a review", Green S., Somers F., May 1997, on-line at
   http://www.cs.tcd.ie/research_groups/aig/iag/toplevel2.html
2. "FIPA, the Foundation of Intelligent and Physical Agents", home-page at
   http://www.fipa.org
3. "Speech Acts", Searle J.R., Cambridge University Press, 1969.
4. "Extensible Markup Language (XML)", W3C Recommendation, February 1998, on-line at
   http://www.w3.org/TR/1998/REC-xml-19980210
5. "Document Object Model", on-line at http://www.w3.org/DOM/
6. "Knowledge Interchange Format Specification", Genesereth, M. R., working draft for an
   American National Standard, March 1995, on-line at
   http://logic.stanford.edu/kif/specification.html
7. "Ontology Markup Language", R. Kent, on-line at
   http://asimov.eecs.wsu.edu/WAVE/Ontologies/OML/OML-DTD.html
8. "Conceptual Knowledge Markup Language", R. Kent, on-line at
   http://asimov.eecs.wsu.edu/WAVE/Ontologies/CKML/CKML-DTD.html
9. "Resource Description Framework (RDF), Data Model and Syntax", W3C Working Draft,
   October 1998, on-line at http://www.w3.org/TR/WD-rdf-syntax
10. "RDF Schema (RDF)", W3C Working Draft, October 1998, on-line at
    http://www.w3.org/TR/WD-rdf-schema
11. "Grasshopper 1.2", IKV++, on-line at http://www.ikv.de/
12. "SiRPAC - Simple RDF Parser & Compiler", J.Saarela, W3C Implementation, 10 March
    1999, on-line at http://www.w3.org/RDF/Implementations/SiRPAC/
13. "IBM XML4J, SAX compliant XML Parser", on-line at
    http://www.alphaworks.ibm.com/formula/xml
14. "XSchema v1.0 specification", on-line at http://www.simonstl.com/xschema/
15. "Document Content Description", W3C Submission, 31 July 1998, on-line at
    http://www.w3.org/TR/NOTE-dcd

Mobility Support with a Mobile Agent System

Anthony Sang-Bum Park and Steffen Lipperts

Aachen University of Technology (RWTH)
Department of Computer Science (i4), 52056 Aachen, Germany
{park,lipperts}@i4.informatik.rwth-aachen.de

Birgit Kreller and Björn Schiemann

Siemens AG, Corporate Technology,
Dept. ZFE T SN1, 81739 Munich, Germany
{birgit.kreller,bjoern.schiemann}@mchp.siemens.de

Abstract. This paper describes the first results and the architectural design decisions of the AMASE project, which addresses the agent technology task of the ACTS third call. The main research interest of the project is to provide the user mobility as a benefit of a mobile agent system. Therefore, the focus in this paper is on the design of a mobile agent system and its environment that meets the demand of high mobility in heterogeneous wireless and fixed networks. We propose a modular agent system to support adaptation and scaling of the system to different devices and network environments. Mobile agents are suitable excellently to realize the user mobility with their capability to operate disconnected. Additional components like an agent launcher and an agent directory are necessary to ensure the autonomous work of mobile agents. First measurements done within a testbed of such a mobile agent system stress the feasibility of the proposed architecture.

Introduction

The explosive development of wireless access networks towards IMT-2000/UMTS and the ever-increasing popularity of the Internet has lead to a growing interest in merging telecommunication and Internet services access. Concerning IMT-2000 [PC97] and UMTS [PC98] in Europe there is a need to support mobile users with any kind of services that can be accessed and used offline. Major efforts have gone into realizing both base technologies for wireless communications, and higher level protocols up to application protocols. However, multimedia applications and services using these new digital cellular systems have not yet been investigated in the same way. Providing mobile users with a higher degree of flexibility and efficiency requires an environment that performs much of the work on the server in the fixed network. In this context the most interesting technology is referred to as mobile agents. The mobile agent approach helps to reduce required network resources and thus supports systems that do not have permanent network connections, such as mobile computers. Mobility and autonomy allow mobile agents to move from their point of origin into a network and continue to operate even if the originating device is temporarily or permanently disconnected from the network. Mobile agents thus can operate in a disconnected way not only for a short timespan, but even satisfy redefined goals without further intervention of users on the move.

The AMASE project investigates and supports user mobility with a mobile agent system [Pa98]. By today, a number of mobile agent systems have emerged which address important issues of mobile agent technology, e.g. migration [RoPo97], collaboration [Mi97], management [Lip98, Cam97, Bie97], and CORBA-integration [Voy, Gra]. Our main focus is on the usage of mobile agents in wireless environments. Therefore, we design, specify, and implement an agent environment taking into account the requirements of an extensive user mobility. In order to run the agent system on small, heterogeneous end devices, the scalability is a crucial issue. We therefore provide a scalable agent system which can be downsized to a light-weight platform. At the same time, the system is to provide a cost-effective adaptability to wireless network characteristics and UMTS service capabilities. The architecture of the mobile agent system caters for a variety of mobile terminals designed for different requirements, including a simple mobile phone, a Personal Digital Assistant (PDA) or a high-end multimedia notebook.

An agent moving through this mobile agent environment is understood as a program that takes over some/most of the users', applications', or even other agents' tasks. If necessary, agents communicate with each other and/or the environment. They are able to move from host to host and launch new agents (during their migration) to accomplish their task. In the AMASE agent environment, mobile agents roam the network, offering and using services. This capability enables new types of distributed applications with a distinguished set of features. These applications can run over heterogeneous networks and on heterogeneous end systems. They can be adapted on the fly by the user or the service provider, can autonomously cope with a range of different situations, and can flexibly exploit the services available in the network. The project uses the mobile agent technology to provide access to information services for mobile users. The envisaged scenario for this service includes:

- Highly mobile users, i.e. those who are commuting on a daily basis during their office and leisure time.
- Equally mobile terminals, including PDAs and mobile notebooks.
- A wide range of networks including but not limited to GSM networks (using SMS, GPRS, general data connections or future UMTS networks if available), and (wireless) local area networks.

One essential decision made with the AMASE system is to distinguish between mobile and system agents. System agents have nearly unrestricted access to local resources and therefore remain at one location and do not have the opportunity to move between different agent systems. Usually they offer services to the mobile agents, and they are completely implemented in Java or simply as a wrapper to already existing applications. Decisive for this differentiation is the security aspect, because mobile agents come to execution in various agent systems and they hence cannot be considered trustworthy. Once a mobile agent enters an agent system, it can only process its internal data and call services controlled by the agent system. The services a mobile agent can rely on while roaming through the agent network become fundamental. It is very important for mobile agents to find services that help to complete their tasks. Therefore, services are always dynamically looked up and invoked through the use of the so-called service center. The service center (SC) takes care of the administration of local services and provides a trading mechanism for remote services. For the service trading, a central repository for all available services is needed. We use an LDAP (lightweight directory access protocol) directory server for that purpose, due to the advantage of providing a distributed, standardized, and publicly accessible database.

The whole AMASE agent system is implemented as a Java library package that consists of standard components that will be described in detail in the next section. First, we will give a detailed description of the agent environment and the agent system. In doing so we will focus on the various inter-agent communication forms. Then we will discuss in more detail the service and agent management proposed by the AMASE mobile agent system. We conclude with the summary and an outlook on future work.

AMASE Environment

The basic approach adopted by AMASE is to take an existing mobile agent system and to enhance it according to the requirements of such a highly mobile environment. The mobile agent system provides a distributed agent environment which comprises several nodes, each running the AMASE agent system. Figure 1 depicts the agent environment that covers all agent systems and the agent directory (AD).

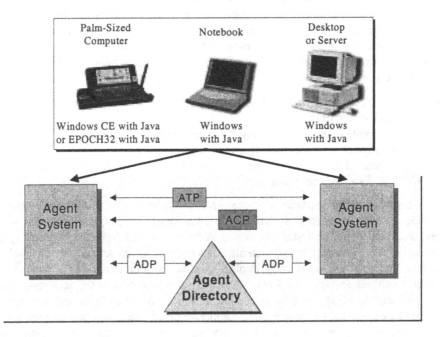

Fig. 1. The AMASE environment

The agent system is scalable and will be implemented during the project period also as a Personal Java version thus it fits in palm-sized computer supported with Windows CE or EPOCH32 and Java as well as notebooks and Unix Server. An AMASE system consists of two main components:

1. The agent system (AS) providing a runtime environment for co-operative mobile agents. For instance, it allows agents to migrate from one agent system to another, to access services available in the network, and to communicate with other agents. A supporting service center (SC) alleviates the task of locating and accessing services as well as agents.
2. The underlying communication facilities (CF) give the AS and thereby the agents access to a broad range of current networks. This can be any bearer service with access to the Internet, e.g. GSM for mobile users, which includes services like SMS or FAX.

At the end of the AMASE project the applicability, flexibility, and performance of the AMASE approach will be demonstrated by means of an application of the banking and financial sector. The next section will outline the system architecture in more detail.

2.1 Agent System Architecture

The agent system is entirely written in Java and among others offers a front-end allowing the administration of agent systems either locally or remotely. Furthermore, an agent system launcher supports loading a scaled version of the AS into a device and executing it on different Java Virtual Machines (JVM) of e.g. JDK or Personal Java. The launcher closely co-operates with a unit for agent system software update allowing to upgrade the AS's software at start-up time or upon explicit request. For applications there is an agent launcher, which allows a convenient and browser-like launching of agent-based applications by hiding all the Java- and agent system specific components.

The core of the AS is the agent manager (AM), giving mobile agents access to the application-specific part of the AS's functionality through the agent API. By means of its communication manager (CM), the AS interfaces to the communication facilities which in turn connects to the available networks and their bearer services or geographic location information as with UMTS. The protocol handlers of the CM establish the protocols used for inter-agent communication, for agent migration, and for accessing the agent directory by means of the service center.

Before both the AM and the CM are discussed in detail, the additional components they rely on are outlined. First of all there is the persistent storage located in the persistent memory area of the underlying device. This component is needed for saving agents and the agent system's state and for configuration issues that are required for consuming work at a later point in time. The user manager and the security manager establish a basic user management, which allows the enforcement of access policies for agent migration and resources. An additional resource manager provides information on the utilization of device resources (e.g. memory or agent population) and also has an impact on migration and access policies. Thus, incoming agents may be delayed until sufficient resources are available. Finally, a component for dynamic update of the agent software is embedded for versioning and updating of agent classes. The agent manager is responsible for controlling the agent population. It allows launching and terminating agents and provides agents the functionality to migrate, communicate, and access services. Two basic agent types are distinguished: mobile agents and system agents. Mobile agents may be created by users or by applications, and they are capable of roaming within a network. However, like applets they are not allowed to access system resources for security reasons. Usually these agents interact with the user for receiving an initial configuration, before they are launched into the network. At some point in time, an agent can return to its origin, but it does not have to. This enables the user to perform disconnected operations without a continuos network connection. This idea is further enhanced by an application-specific reachability management which tries to deliver results to a user over the information channel he/she is most likely to be available. For instance, if the user switched off the agent enabled PDA, a FAX or SMS may be used as an alternative. Besides the mobile agents, there are special system agents that do not migrate between agent systems. Since their origin is well known, they may be granted access to system resources. This enables them to become a mediator between the system and those mobile agents which want to access system resources and services. Besides this basic distinction, the agent manager co-operates with the user manager and the resource manager which allows assigning detailed access rights to mobile agents. Both agent types are maintained separately by the agent manager, which allows a clearly defined type-dependent treatment, e.g. in case of a shutdown.

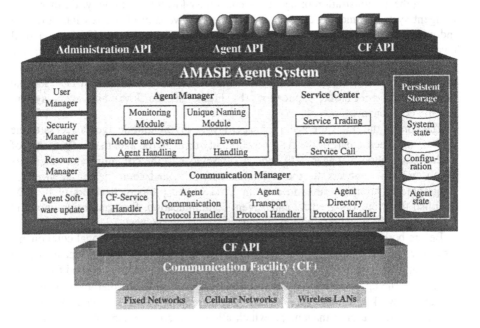

Fig. 2. Detailed architecture of the agent system

As can be seen from figure 2, the agent manager provides three interfaces which give access to its functionality. First there is an administration API which allows configuring the agent system, e.g. concerning the maximum size of the agent population or timeouts used for handling and optimizing network connections. This includes default servers and directories as they are later introduced for the service center. This administration API is not available to applications, i.e. mobile agents, and will only be used by external applications like the administration front-end or the agent system launcher.

The agent API is dedicated to agent-based applications and enables agents to perform the following actions:

- Control the agent lifecycle including creation, launching of other agents as well as their own termination and migration.
- Handle events via an event queue located in an agent's main thread.
- Communicate synchronously or asynchronously with other agents.
- Access services provided by different bearers. This is enabled through trader-based commu-ni-cation which relies on the directories and mechanisms introduced in the following.
- Involve the security manager into application- or user-specific security policy enforcement.
- Query the resource manager for information about the underlying agent system and device.
- Access the persistent storage for saving the agent state.

Finally, the CF API is not system specific but enables system agents a transparent access to UMTS services like geographic location of mobile devices or user profiles.

2.2 Inter-Agent Communicatio

The communication manager (CM) is responsible to connect the entire agent system to the communication facilities (CF) which in turn connect a device to the available networks and the outer world. The CM listens on pre-configured ports at sockets provided by the CF in order to receive incoming messages and agents which will then be dispatched and handled by the agent manager. It is responsible for converting Java objects into byte streams, and it is also involved in synchronous communication that requires temporal suspension of agents. By the means of the CM and its dedicated protocol handlers, the following protocols are established over the network connections:

- A CF service protocol is used for accessing system functionality and telecommunication services like FAX and SMS or UMTS-based services. This protocol will not be publicly available, thus only system agents may use it.
- An agent directory protocol is used for accessing all the directories and functionality provided by the service center. Basically, the related protocol handler performs a wrapping of the well-known LDAP protocol used for the service center.

- The agent transfer protocol is used when agents have to be migrated between agent systems.
- Finally, the agent communication protocol is established for synchronous and asynchronous inter-agent communi-cation mechanisms.

Together with the CF, the CM optimizes communication and connection handling. For instance, the used protocols take network and device characteristics as well as quality of service informa-tion into consideration. Furthermore, con-nections will be physically closed by means of timeouts and kept open only virtually. This is fully transparent to agents, helps to save connection costs, and further supports disconnected operations of mobile devices and their applications. By means of the agent system's communication manager, its protocol handlers, and the underlying communication facilities, the following basic communication mechanisms are implemented:

- asynchronous one-way agent-to-agent messages;
- synchronous two-way agent-to-agent messages which are based on remote procedure call mecha-nisms (RPC).

A message consists of the message content and the destination address comprising the destination system's name as well as a name pattern denoting the receiving agent. As far as synchronous mechanisms are concerned, the name of a procedure/method to call at the destination will also be submitted. Synchronous communication returns a result, which may be some data or an exception in case of encountered errors. Within our agent system all communication, both on application and system level, is based on these basic inter-agent communication mechanisms. They rely on appropriate system agents and even allow handling agent migration via the same mechanisms.

2.2.1 Asynchronous Inter-Agent Communication

This communication style enables agents to send messages without having to wait for a result, because the receiver is not expected to provide it. However, the sender can never be sure that its message really reached its destination, which is due to the mobility of agents. For instance, the destination agent may have migrated in the time interval lying between the sender locating the receiver and the sending of the message. Only in case of communication errors (e.g. un-availability of net-works), the sender will be notified. However, this is not a reliable information, because there might be further impacts like broken network connections preventing messages from being delivered. Irrespective of this drawback, the concept of asynchro-nous communication is valuable for mobile agents for informa-tion dissemination purposes including multi-casting.

Before the sending agent may send its message, it has to know the receiver's address information which can be obtained from the service center's agent directory. The sender just submits this information as well as the message content to its send-Message method, and from there on the agent system will handle all of the communication. Thus the sending agent may continue execution and try to catch an exception indicating the potential failure of its communication attempt. The underlying socket-based agent communication protocol distinguishes between local and remote connections in order to save resources. On the receiver side, messages may be filtered. Afterwards the selected messages will either be inserted into the receiving agent's event queue or handled directly.

Additionally, the agent system provides a blackboard for local asynchronous communication. Such a blackboard is a data area where agents can leave information entities that may be read and removed by other agents at a later point in time and under configurable access restrictions. The entities are marked with a time to life variable and will be removed automatically from the board if they expire.

Another asynchronous inter-agent communication is offered by the so-called post-box. A post-box is a message queue which belongs only to a single agent that solely has the access rights. All senders aware of the post-box location are allowed to deliver messages to it. The owner can read the messages in a predefined order (e.g. FIFO or LIFO).

2.2.2 Synchronous Inter-Agent Communication

The synchronous inter-agent communication via an agent's request method is different from the asynchronous one in the sense that the sending agent will be halted until the receiving agent provides the results for the communication request. As with the asynchronous form, the sending agent at first must obtain the receiver's address information which will later direct the message to its destination. If the receiver resides within the same agent system, then the request will be handled locally. Otherwise the sending agent system will open a socket connection. On the receiving agent system, a socket listener will take the incoming request and deliver it to a dedicated thread of the receiver agent if it passes filtering. There the request will be handled directly or via its agent's event queue in order to reach the method specified within the original request call. Here too, the results that must be delivered back via sendResultMsg to the sender will be collected, as they become available. When the result message (which may also contain error notifications) arrives at the sending agent system, the sending agent will be woken up in order to resume execution and evaluate the results of its request call. Synchronous communication is supported between those agent systems which can establish a direct socket connection.

2.2.3 Support of Disconnected Operations

It is likely that mobile users activate agents through agent based applications or simply invoke agents through small mobile devices which are not necessarily agent enabled. Therefore, the AMASE system supports PDAs with GSM connection as well as multimedia notebooks. To meet these requirements, we consider the limitations of the calling machine. There are two main restrictions: the ability of the machine to process Java Bytecode or some other high-level language, and the machine's interconnection with the fixed agent network. The former determines the type of launcher, the latter the method, how results are returned to the machine or the user. In AMASE, we will offer two ways to launch mobile agents:

1. a scalable agent system with only the core components for machines which are able to process Java Bytecode and
2. an agent launcher protocol for very small devices without a Java Virtual Machine.

The core agent system implements the components necessary to execute and send an agent to the next agent system and optionally to accept incoming agents. The agent launcher protocol allows applications to hand over information about the class files of a mobile agent and how the results are to be returned to a dedicated agent system in the fixed network. At the corresponding agent system, the identified mobile agent is created and can accomplish its task.

There are several ways to return results to the originator once the agent has finished a task. First of all, using services like e-mail or fax can help, but it is also possible to return the entire mobile agent to the minimal agent system. In the latter case, some difficulties must be resolved. If the launching agent system is not reachable it has to be informed (again via a voice call or similar services) that the agent has fulfilled its task and wants to return. Then the mobile agent has to wait until the disconnected agent system is reachable or it uses some kind of agent-hosting service and is suspended until the disconnected agent system picks it up.

Service and Mobile Agent Management

The previous chapters have introduced the architecture of the AMASE system and have pointed out the mechanisms provided for communication of system agents and mobile agents. It is the communication of the latter which introduces a new and important issue to the process of inter-agent communication. Since mobile agents are capable of migrating and since this is a process which can take place at any point in time, it is necessary to introduce a mechanism to determine an agent's current location. While this is not the case with communication based on blackboards or post-boxes, it is inevitable to be initiated with direct communication of agents. For this purpose, the Mobile Agent System Interoperability Facility (MASIF) [OMG] specifies a MAFFinder, an abstract facility for mobile agent localization. The MAFFinder is abstract in that it does not specify how the agents are to be localized, merely the presence of such a facility is required. In the following, the concepts for mobile agent management, in this case agent localization, motivated in MASIF are presented, namely broadcast, forwarding, and directory service/home registry. The concepts are evaluated and the approach chosen and implemented by AMASE is discussed in detail.

As a first localization concept, broadcasts can be deployed to determine the whereabouts of agents. If a communication is to be initiated with a mobile agent, a broadcast can be issued to all agent systems, and the system currently hosting the agent will reply with its reference. This approach holds the advantage that after an agent's creation, there is no need for any kind of interaction to keep track of its position and the mobile agent can freely migrate and use services. It is not until the actual moment of establishing a communication that the agent's location needs to be determined. However, this approach also has grave drawbacks. There is the danger of flooding a system with broadcast messages if the agent localization is required more frequently, which will lead to high network loads and system utilization. More importantly, this approach does not scale in open systems, as broadcast messages will not be efficiently possible.

Another approach therefore is for the agent to leave a trail of its migration path. This can be done with forwarding. Each time a migration takes place, a forwarder is installed, pointing at the agent's new location. If an agent reaches a location it has visited before, it will remove the circle of forwarders and continue from there. In case of an agent lookup, this trail can then be followed to the current agent location. The main advantage of this approach lies in the guided search for the agent, which makes a global search over all systems unnecessary. However, this comes at the price of possibly long trails for the agents littering the agent environment. In addition, the agent lookup times can grow considerably if the paths grow long. Most importantly, the mechanism is not reliable, as it cannot guarantee that the agent will actually be found, because a break at any point at the chain of forwarders will result in the agent being no longer to be found.

While both combinations of these two approaches, and probabilistic models of the agent migration process [ChLe97] can also be regarded as promising for mobile agent lookup, the AMASE system relies on an entirely different approach which is introduced a service center and is based on a directory service. This approach makes use of the general mobile agent execution cycle. Due to the costs of agent migration [Lip98], mobile agents are restricted in their size and thus in their complexity. This implies that mobile agents will have to use services to execute all the tasks required. Consequently, the agents will need to contact a facility in the agent system which provides e.g. a naming or trading service and which will pass information on the whereabouts of services. This principle has been transferred to mobile agent lookup as part of the inter-agent communication. The service center, the AMASE facility containing a trader for service lookup [PaLi98], has been extended by a component to keep track of the mobile agents, as shown in figure 3.

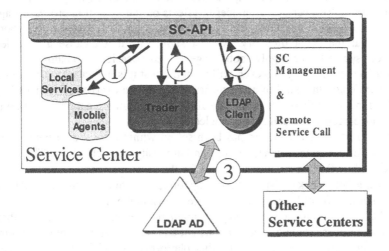

Fig. 3. Architecture of the service center

The process of agent localization is now shifted from the point, where a communication is to be established, to the migration process. Whenever a mobile agent is migrating to another host, its position will be updated in the service center. The actual storing of the agent location is done in the agent directory (AD). This is implemented as an LDAP server [Mic96], with the SC thus holding an LDAP client for accessing the agent directory. The SC is in charge of the administration of local services and provides a trading mechanism for remote services. For the service trading, a central repository (this can be a common database, the CORBA naming service or an X.500 directory service) is needed for all services available. The SC has a common application-programming interface (SC-API) for all mobile agents and provides functions necessary to access the AD, suitable for both mobile and system agents. The SC-API is completely separated from the AD. Thus, the SC is independent and acts as a mediator between mobile agents, service agents, and the AD. Consequently, the SC plays a central role and provides agent administration as well as system administration, i.e. registering and de-registering of services, agents, and systems, and offers an API to service and mobile agents acting as a trader. As shown in figure 3, all components communicate through well-defined interfaces, and therefore the integration of new trading mechanisms or directory services is fairly easy and does not affect other components of the SC.

This approach of deploying an agent directory holds a number of advantages. First, there is no need for localizing a mobile agent if it is to be contacted, as its current position is always known by the service center. Thus, lookup delays are avoided. Moreover, the agent lookup mechanisms is reliable, as the update of the agent position is embedded into the agent migration process, i.e. a migration is not completed before the update has been executed. Consequently, mobile agents will never be lost track of. There will also be no message bursts caused by agent localization. Finally, the agent directory concept allows a seamless integration with the facilities required for localizing services for mobile agent usage. Deploying an agent directory, however, has a number of effects on the migration process of agents. Whenever an agent is migrating, the agent system needs to update the corresponding entry held at the agent directory. At first sight, this might cause a considerable overhead, especially for highly mobile agents and occasional lookups. The results presented in the following however, will point out that it is in fact very reasonable and efficient to deploy this kind of agent management, as the overhead introduced by the updates in the agent directory is considerably smaller than the actual migration times of the mobile agents, and can thus be neglected.

A testbed of the agent system has been set up within a local area network at the department of computer science of Aachen University of Technology. The switched Ethernet with 100 Mbit/sec was sufficient for the local area network testbed and did not caused a bottleneck. First measurements done with the standalone LDAP server (SLAPD) of the OpenLDAP organization are shown in the figure 4. The printed mean values have been achieved by 100 measurements per request and show that the response time of the lookup is nearly constant and remains stable even if the number of migrating agents increases. The hosting platforms are SUN Ultra 1 workstations (167 MHz) equipped with 128 MB RAM. Considering only the look-up and migration time a first statement that can be made is that registering and de-registering of mobile agents remains constant and is not necessarily the delaying component of an agent migration. It is obviously that replacing the agent directory with a faster database or directory can decreases the tracking of mobile agents if a bottleneck does occur. In our measurements the average lookup time amounts to 60 ms and mainly represents the time needed for network set-up to the agent directory, the time for data exchange and the search process in the LDAP server is negligibly small. Conceptually, real systems like GSM have shown that home and visitor registers are capable of managing millions of users. A more hierarchically organized structure for the user management proposed with UMTS would allow a more distributed and decentralized organization, which will increase the number of manageable users [KIS98].

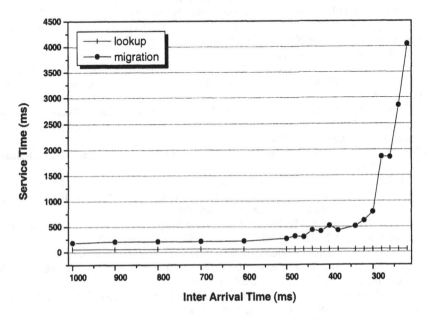

Fig. 4. Lookup and migration time depending on agent system load

Summary and Outlook

Agents are a promising new technology. Especially in conjunction with wireless access networks they offer exciting new possibilities to support the mobile user. This paper summarized the rationale behind, and the basic architecture of the AMASE project, giving some insights into AMASE's technical approach. The focus presented in this paper was on mobility support with the AMASE mobile agent system. First measurements have shown that the lookup time remains nearly constant, independent of the requesting rate of the mobile agents. The usage of a central inquiry register is still the simplest and most common method to keep track of agent mobility and service locations. An efficient and distributed data structure for the lookup service is undoubtedly important, but real systems used in current cellular systems show that this is a solvable engineering task. The support of mobile users with the AMASE agent system will be demonstrated and verified in an agent-based application, which will show features in the banking and financial service area. The experiences and results of the project will be contributed to the FIPA standard.

References

[Bie97]   Bieszczad: Advanced Network Management in the Network Management Perpetuum Mobile Procura Project. SCE Technical Report SCE-97-07. March 1997.

[Cam97] Campeau: Fourth Year Project Final Report: Managing Networks with Mobile Code. Fourth Year Project Final Report, April 1997.

[ChLe97] W. Chen, C. Leng: A Novel Mobile Agent Search Algorithm. Proceedings of First International Workshop Mobile Agents '97, Berlin, Germany, April 1997.

[Gra]   Grasshopper homepage: www.ikv.de/products/grasshopper

[KIS98]   A. Küpper, F. Imhoff, S. Hoff: Evaluation of Agent Concepts for Service Providing in 3rd Generation Mobile Networks. 3rd Workshop on Personal Wireless Communications, PWC'98, Tokyo, Japan, April 1998.

[Lip98]   S. Lipperts: CORBA for Inter-Agent Communication of Management Information. 5th International Workshop on Mobile Multimedia Communication, Berlin, Germany, October 1998.

[Mi97]   Mitsubishi Electric Information Technology Center America: Concordia - An Infrastructure for Collaborating Mobile Agents. Available at: www.meitca.com/HSL/, Horizon Systems Laboratory, 1997

[Mic96]   The University of Michigan: The SLDAP an SLURPD Administrator Guide Release 3.3, April 1996.

[ODP97] ODP Trading Functions: Specification, Draft Rec. X.950-1 ISO/IEC 13235-1, 1997

[OMG]   The Object Management Group (OMG): Mobile Agent System Interoperability Facilities Specification 98. TC Document, March 1998.

[Pa98]   A.S. Park: The AMASE Project: Agent based Mobile Access to Information Services. Workshop on Agent Technology 98, Baltimore, USA, November 1998.

[PaLi98] A.S. Park, S. Lipperts: Prototype Approaches to a Mobile Agent Service Trader. 4th International Symposium on Interworking "Interoperability of networks for interoperable services", Interworking98, Ottawa, Canada, July 1998.

[PC97]   IEEE Personal Communications, IMT-2000: Standards Efforts of the ITU, IEEE Communications Society, Vol. 4 No. 4, August 1997.

[PC98]   IEEE Personal Communications, Paving the Way to Third Generation Mobile Systems in Europe, IEEE Communications Society, Vol. 2 No. 2, April 1998.

[RoPo97] Kurt Rothermel and Radu Popescu-Zeletin (eds.): *Mobile Agents*, Lecture Notes in Computer Science 1219, 1997.

[Voy]   Voyager homepage: www.objectspace.com/products/voyager

Software Agents for Enhancing Messaging in a Universal Personal Numbering Service

Ramiro Liscano, Katherine Baker, and Roger Impey

Institute for Information Technology, National Research Council
1500 Montreal Road, Ottawa, Ontario, Canada, K1A 0R6
{Ramiro.Liscano, Katherine.Baker, Roger.Impey}@iit.nrc.ca
http://www.iit.nrc.ca/SPIN_public

Abstract. This paper introduces the Universal Personal Numbering System (UPN), an application that works with a Personal Agent Mobility Management System (PAMMS) to manage multiple communications devices. It allows messages to reach the user on different phones or messaging devices using one single number. This paper includes details on the design of the UPN application and how it works with a UPN service adapter agent to notify non-telephone devices of incoming calls and await instructions from the user. Additionally, there is a description of the interface between the UPN application and a PC-based switch using the TAPI protocol.

1 Introduction

Over the past several years, increasingly diverse types of personal communication devices and services have become available to users. Multiple telephones, cellular phones, pagers, two-way pagers and personal digital assistants (PDAs), and services such as text-messaging, call forwarding, and voice mail have allowed users to have instantaneous access to their personal communications from multiple locations.

The rapid development in communications services has been driven by a number of factors including the worldwide de-regulation of the telecommunications industry, the introduction of affordable digital wireless services, and the ever-increasing availability of information on the Internet. Currently, there are four commonly used networks for communication:

1. The Public Switch Telephone Network (PSTN), used primarily for voice transmission and, increasingly, data services via modem, ISDN, and xDSL connections.
2. The wireless PCS-oriented networks offering voice services and, by allocating particular voices channels for data, data services.
3. Wireless data networks, designed in particular for data transmissions, have focussed in the past years on vertical market applications for two-way communications and one-way paging for the horizontal market. Access to the Internet is now driving most horizontal two-way wireless data communication applications.

4. The Internet, a packet based network offering electronic mail messaging, has now become the driving factor for the integration of most voice and data communication services. Internet-based universal messaging systems now store asynchronous communications messages like FAX, e-mail, and voice.

This growth and diversity in communications services has resulted in users having several devices each with its own unique message transfer protocol and addressing identification. For the horizontal market, the two principal addressing schemes for communicating with users are telephone numbers and e-mail addresses. There have been several atempts in the research community to try and define a universal personal number such as Eckardt's et al [1] approach for a universal personal addressing system based on X.500 and X.700 standards, and Schulzrinne's [2] universal personal telecommunication system based on existing Internet protocols for video-phone communications. These attempts did not encompass the telephony world (telephones and cell phones) and restricted the user to desktop applications. The reality of the current situation is that numbers are easier to use for telephone interfaces and names for text input. These attempts towards a Universal Personal Number (UPN) are primarily driven by Europe's third-generation mobile computing initiative known as The Universal Mobile Telephone System (UMTS) [10] where a single system is desired for residential, office, cellular, and satellite environments. Also the convergence of telecommunications and information technology [13] requires the management of diverse communication networks.

With this in mind, this paper introduces the concept of Universal Personal Number, a single telephone number that, with the help of software agents, accesses multiple communication devices. The UPN application works within the framework of a Personal Agent Mobility Management System (PAMMS), an agent-based system that manages and acts on user preferences to manage messages in various forms. The messages can exist in different modes (e.g. email, text, voice, FAX etc.), be directed through different networks (e.g. telephony, wireless, Internet etc.) and be passed using different protocols (e.g. HTTP, POP, TAPI etc.).

Though this application could have been developed without the use of agents, this agent-centered design has several useful features beyond simply using distributed objects. These include:

- A modular architecture based on message passing among the agents where is agent is implemented as a standalone process with its own reasoning engine and knowledge base.
- A common language of communication is used among the agents based on speech act theory. Agent's can query each other's knowledge using standard communication performatives rather than published method calls.
- Each agent can be dynamically programmed since the Agent Communication Language (ACL) supports the manipulation of an agent's knowledge.
- Reusable code since all agents are cast from one mold and easily customized with their own reasoning and knowledge engine.

2 The Personal Agent Mobility Management System (PAMMS)

In order to achieve a flexible, scalable and adaptable system to manage personal communications, we introduce a seamless messaging system based on the use of software agents [3]. The Personal Agent Mobility Management System (PAMMS) extends previous efforts in agent-based personal communications systems such as those presented in the DUET [4] project and and the PCSS TINA-C auxiliary project [11]. It utilizes personal communications agents, service adapter agents and service-specific gateways to accomplish interception, filtering and delivery of multi-modal messages. The agents use the Java Expert System Shell (JESS) [5] for internal reasoning and knowledge handling and the Knowledge Query and Manipulation Language (KQML) [6] for inter-agent communication.

2.1 The PAMMS Architecture

The Personal Communication Agent (PCA) is the central agent in the PAMMS architecture. It uses service adapter agents, resource managers and service gateways to accomplish seamless message transfers. PAMMS can be classified as a set of networked agents according to the taxonomy defined by Magedanz et al [12]. Network agents are designed to access both local and remote resources and act as personal assistants. The agents hide the complexity of the network and resources from the user and configuring services on the user's behalf. This particular PAMMS implementation is a classical hierarchical structure with the PCA at the top of the hierarchy, the service adapter agents as the middleware software components, and the gateways as the interfaces to services available on different transport networks. The hierarchy is determined primarily based on the roles of the agents. Figure 1 shows a conceptual diagram of the system components.

The PCA supervises the service adapter agents and makes decisions based on events received from the adapters. It subsequently issues requests to the adapters to pass on to the gateways or programs the adapters to perform particular functions on behalf of the PCA. They represent the user's choices and desires on how to manage its communication environment. They also are responsible for classifying and performing actions on all incoming messages.

The service adapter agents can be implemented as either wrappers to the gateways that only translate requests between the gateways and the PCA or as intelligent agents that possess the logic required to process an event from the gateways and react on behalf of the user. The UPN service adapter agent is an implementation of the latter type where the adapters are programmed to respond on behalf of the user.

The PAMMS architecture can also integrate the use of translation services such as optical character recognition and text-to-speech converters. The service adapter agents utilize these content converters to translate messages into different mediums.

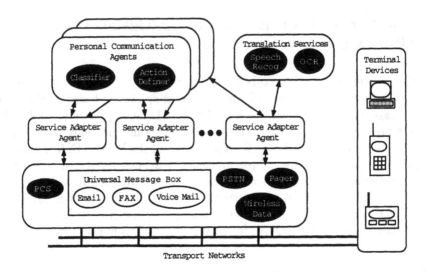

Fig. 1. Personal Agent Mobility Management System Architecture

The PAMMS system is designed primarily to notify the user of incoming communications and to establish a connection to the user for isochronous forms (i.e. telephony) of communications. Alternatively, the user can establish a connection to his or her messaging environment using a conventional browser or dial-up approach. The aim of the PAMMS architecture is to provide an agents-based system model for the notification of incoming messages, either connectionless or connection oriented, to the user on their preferred device and customized to the user's specifications.

Notification and the establishment of a connection are two features that are fundamental to all communication mediums. In the current generation of communications, these two functions are tied together into one application and integrated into one device. In future generation systems, this does not necessarily have to be the case. With today's sophisticated messaging environments, notification, connection, and retrieval can be separated to take full advantage of the different modes of communication available. For example, it is possible to receive notification on one device and request to have the message routed to another device. This type of functionality is readily available in advanced "Call Centers" where call notification can occur on a service agent's desktop and call redirection can occur at the discretion of the agent (in this case, a person as opposed to a software agent). This example can also be demonstrated for asynchronous communication where it is possible to receive notification of an e-mail on a pager or phone and retrieve the e-mail using a desktop application.

2.2 Agent communication model

Communication among the agents is based on using the Knowledge Query and Manipulation Language (KQML) performatives defined by Yannis and Finin [6].

KQML has a well- defined structure for agent to agent communication with fields that define the recipient agent, sender agent, ontology, language, message identifier, and content of the message. In particular, the language and the ontology fields are unique over other common communication protocols. Each agent in PAAMS is programmed with a unique ontology using the Java Expert System Shell (JESS).

In one model of PAMMS, direct communication between the service adapter agents is not necessary as the common flow of communication is hierarchical, passing all messages to the PCA which redirects them to the appropriate adapter agent. This centralized model of communication has been implemented for several PAMMS examples [7]. It is the simplest to configure since all of the service adapter agents are independent and do not need to share knowledge with each other. A second, more distributed model of PAMMS utilized by the UPN application allows for direct communication between the service adapter agents. It requires a mechanism within the adapter to allow programming of knowledge in another service adapter agent. The advantage of this approach over a centralized model is that there is a direct path of communication among the agents removing possible bottlenecks. This leads to a more distributed system model.

2.3 Anatomy of a PAMMS agent

Each PAMMS software agent contains three basic modules: a communications module, a reasoning component, and a knowledge repository. Figure 2 depicts the structure of a typical agent. This particular agent contains two communications modules, one for agent to agent communications and the other for communicating to non-agent processes. The latter module is tailored to specific application programming interfaces offered by manufacturers to access particular communication services or networks. It is, therefore, more of a channel for querying or controlling services on particular networks rather than a communications channel with the agent. This is quite different from the agent communication module where it is possible to query the agent.

Figure 2 shows the paradigms used to implement each particular module. Three commonly known methodologies were adopted in the PAMMS agents. JESS is used for coding and querying the agent's private knowledge base and KQML is used for inter-agent communication. The agent itself was developed using the Java programming language.

Each PAMMS agent is implemented as an executable object that waits for messages that are sent to it. Each of the modules is designed as a package with their particular objects and methods as interfaces to the package. Within the agent itself, there is a natural flow to the information. All KQML formatted communications relating to the agent are parsed through the agent communication module, translated to JESS and placed in the agent's knowledge base. The reasoning module interacts with this knowledge base and can either add or delete knowledge as well as invoke methods associated with the other communication modules; i.e. to send messages to other agents or to an external gateway.

Fig. 2. Service Adapter Agent Design

Creation of a generic agent is rather simple as it can be constructed from the basic agent objects.

The agent is designed to run in a multi-threaded mode where it is possible to create several concurrent executable reasoning threads that can manage several independent reasoning tracks. This has been exploited in the design of the service adapter agents since they must simultaneously service requests for communications from several PCAs.

2.4 PAMMS Implementation

The agents are implemented as distributed objects that are CORBA enabled and rely on the CORBA communication infrastructure for communications (see figure 3. Wherever possible, communication to the gateways is via the CORBA event channel or standard client-server CORBA IDL requests.

3 Implementing a UPN Service Adapter Agent

3.1 The TAPI Gateway

Gateways in PAMMS provide a means of communicating between the Messaging System and services on other non-PC-based networks. They allow users and agents to monitor these services for activity and initiate actions.

For the system to manipulate telephone call functions, a TAPI Gateway was developed that would interface the telephony world. This application can successfully connect to any switching server that uses Microsoft's Telephony API. In implementation, the switching server took the form of a Mitel MediaPath Server and Telecom Extender. This hardware/software layer interacts with the TAPI

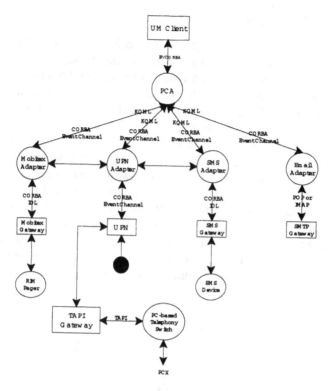

Fig. 3. Detailed Implementation of PAMMS

Gateway via the TAPI layer, an interface provided by Microsoft for Windows systems.

The call model used by TAPI is high level, and does not allow for complete control over calls that more sophisticated protocols allow. TAPI purposely takes care of the low level details and simply informs users of events that have already taken place. This allows applications to remain simple and quick to develop, rather than being riddled with details that are probably unrelated to the application's purpose.

The TAPI Gateway uses the CORBA Event Service to allow client applications to passively (e.g. monitoring call activity) or actively (answering and placing calls) interact with the telephony switching system. It is comprised of a telephony switch monitor and a communications bridge.

When the Gateway is started, the telephony switch monitor establishes a communication link by successfully executing the lineOpen function with the TAPI server. Subsequently, it initializes both a thread to monitor telephony activity and a command executor stack to receive requests and sequentially send instructions to the TAPI server. The communications bridge is responsible for establishing a proprietary event channel using the CORBA Event Service.

3.2 The UPN Service

The Universal Personal Numbering Service works with the TAPI Gateway to re-route calls to specified locations according to user preferences. Each universal personal telephone number can be forwarded to any number of different devices including cellular phones, pagers, home phones, work phones, and voice mail systems. Notification of incoming calls can be received on two-way pagers and short messaging system (SMS) devices.

The UPN System maintains a database of device information and user preferences for each universal personal number, see figure 4. Each entry in the UPN database contains the user's name, the universal personal number, a list of forwarding phone numbers, a selected forwarding and voice mail number, and the number of rings until line termination. It also maintains a list of devices that could be notified in the event of an incoming call, a selected device and a "Check Agents" flag that tells the system to wait for instructions from the notification device before forwarding the call. An activation flag that indicates whether the preference is active or not is maintained for each entry.

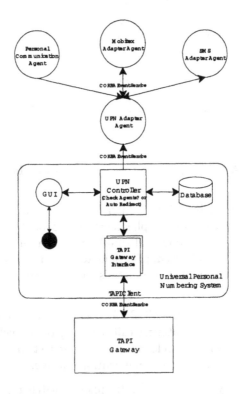

Fig. 4. UPN Architectural Design

When the UPN Service is initialized, the UPN Controller initializes a search on the database, extracts all of the UP numbers and instructs the TAPI Gateway to forward all calls for those numbers to the UPN Controller. Each line is initialized as a consumer object of events produced from the TAPI Gateway.

When a call comes in on one of the activated lines, the TAPI Gateway must check with the UPN database for instructions on where to direct the call. If the "Check Agents" flag is activated, the controller will instruct the UPN Adapter Agent to send a message to the appropriate service adapter agent (e.g. the Mobitex adapter or SMS adapter) and wait to receive a forwarding line number. This reflects the de-centralized model for the storage of knowledge discussed earlier. In the centralized model, the instructions for the Mobitex and SMS adapters would have been passed through the PCA. In this case, the knowledge is handled directly by the UPN service adapter agent with no interference from the PCA.

If the flag is not activated or there is no response from the device, the call is transferred to the selected forwarding number from the database. If the line is busy or there is no answer after a specified number of rings, the call is redirected to the voice mail number. If there is no voice mail, the call is dropped.

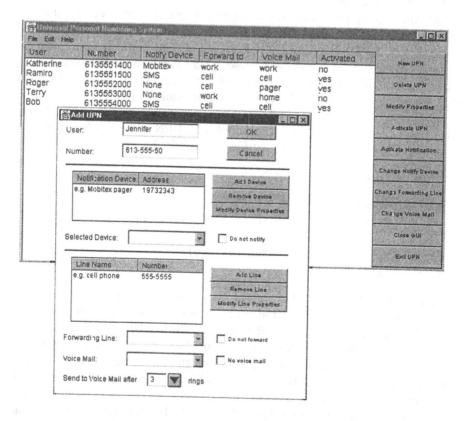

Fig. 5. Graphical User Interface of UPN System

The UPN database can be configured directly by the user using the GUI depicted in figure 5 or by the UPN Service Adapter agent that acts on instructions from the PCA. When a user or agent activates or deactivates a set of preferences, the database monitor is notified and makes the necessary changes in interactions with the TAPI Gateway.

3.3 Programming the UPN Service Adapter

Each service adapter agent in PAMMS is programmed to interact with a particular service by defining a set of JESS rules and knowledge that pertain to that service. The basic particle of information stored in a JESS knowledge base is a fact that identifies some piece of information about the state of the world. JESS is a rule-based decision system developed in Java and based on the CLIPS Expert System.

JESS utilizes a RETE pattern matcher allowing for a variety of approaches in specifying facts in the knowledge base. Most facts though are represented either as attribute value pairs, for example (attribute value), or as an object label with multiple attribute value pairs such as (object-label (attribute1 value1) (attribute2 value2)...).

Attribute value pairs are used to represent simple facts. For example, the ontology that an agent understands is represented as *(ontology-supported UPN)*. Facts that define objects using multiple attribute value pairs are akin to the use of frames in classical AI knowledge representation. For example, the fact *(service-agent (name UPNAgent) (serviceSupported UPN))* specifies a label service-agent that is named UPNAgent and supports a service called UPN.

Reasoning is coded as 'if...then' statements operating on the current knowledge base. The following code is an example of a UPN adapter rule to act upon a user's request to change the forwarding number for his UPN service.

Example of a UPN Adapter Rule

```
(defrule forward-call-to
?fact <- (forward-call (custID ?cID)(forward ?to)(custr ?passWd))
=>
(changeForwardTrans ?cID ?toNum ?passWd)
(retract ?fact))
```

This rule matches on a fact in the knowledge base with the object label *forward-call* and extracts the customer's identification *(cID)*, password *(passWd)*, and forwarding number *(to)*. It subsequently calls a JESS user-function called *changeForwardTrans* to perform a transactional service with the UPN service.

This logic is contained in a JESS startup file for the agent as specified in the constructor method of the agent. This file contains a collection of batch commands that will load a number of both standard and service-specific ontology files. Standard ontology files include a KQML ontology that defines the knowledge required to understand and act on KQML performatives, a security ontology that can manage secure agent to agent communication, a content translation ontology that allows an agent to use the translation services, and a UI ontology that allows for the addition of a user interface to an agent.

Each service adapter agent will load a number of service-specific ontology files. For instance, the UPN agent requires the "UPNAdapter.jess" file which contains all the knowledge rules and facts for the adapter. This is where the *forward-call-to* rule specified above is declared. The agents that will initiate communications with another agent must also contain knowledge about the other agent. This knowledge is stored in a file for the "client" knowledge. For example, any agent needing to configure preferences in the UPN adapter, would need to load a file called "UPNClient.jess". In essence, a client-server model is used for exposing the agents to the knowledge of the other agents.

The client files have rules that are similar than those described for the adapter, except that they request that a KQML message be sent to the appropriate adapter agent. For example, the UPN client rule shown below is the match to the *forward-call-to* adapter rule listed before.

Example of a UPN Client Rule

```
(defrule UPNForwardCallTo
?fact <- (forward-call (custID ?cID)(forward ?to)(custr ?passWd))
=>
(service-agent (name ?name) (serviceSupported UPN))
=>
(KQMLTellFact ?name UPN ?fact)
(retract ?fact))
```

This procedure is akin to exposing the interface of an agent except that this information is not truly an interface; it is the knowledge that will cause the adapter rules to fire and act upon this knowledge. It is not required, however, that all agents be programmed with the knowledge of each other. Only those agents that initiate communications with another agent need to know what knowledge the other agent will react to. In other words, when an agent is behaving as a server to a client agent, it does not need to know the client's interfaces.

Figure 6 shows a simple client and adapter configuration of one agent communicating with one service adapter. This figure also depicts a third file that defines common JESS structures and knowledge that will be used by the both the adapters and the client agents.

A UPN service adapter agent has the same agent structure as others but contains the JESS knowledge to manage the UPN service and communicate with the other agents. This type of configuration lends itself well to the addition of adapters. Unfortunately, this cannot be done dynamically at this point since the JESS files are loaded at run time. Future releases of the PAMMS architecture will consider the actual transfer of this knowledge using the agents themselves rather than the file structures. At this current stage it is also difficult to remove the knowledge associated with a particular ontology since each set of knowledge

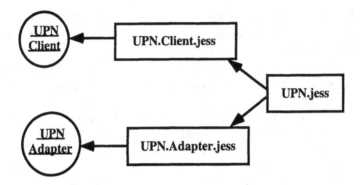

Fig. 6. Example of programming the agents using JESS files

is not truly packaged with its corresponding ontology and therefore cannot be distinguished from other knowledge the agent contains.

4 Use Case

A typical user will possess multiple telephone devices and one non-telephone notification device. They will first set up an account using their PCA. Using the Unified Messaging Client, a GUI provided to load rules into the PCA, they will create a new UPN and provide the relevant information to establish an account. For example, they would provide their name ("Jennifer"), their universal personal number ("613-555-5000"), the notification device and address ("Mobitex pager", "13450983") and a list of phone lines ("cell phone", "work phone", "home phone"). They will then select a default forwarding line ("cell phone") and a default voice mail line ("work phone"). They will also select the preference that their Mobitex pager first be notified before forwarding any calls. The UM Client will program the PCA which will notify the UPN Service Adapter Agent which will create a new record in the UPN database. The UPN Controller will establish a connection with the TAPI Gateway and request that all calls coming in on line "613-555-5000" be answered and wait for instructions from the UPN.

When a call does arrive on this line (from line 613-555-2222), the caller hears a "please wait - redirecting call" message and a notification message is sent out via the UPN Agent. This agent transmits the message to the Mobitex Adapter Agent which works with the Mobitex Gateway to send a text message ("Call from 613-555-2222") to the two-way pager. The user will then have the opportunity to drop the call or redirect the call to a specified number or one of the lines described in their account.

For this example, they will select "redirect to home phone". The call will be transferred to the home phone number from their account. If the line is busy or there is no answer, a new notification will be sent to the Mobitex Gateway. In this case, the line was busy so a new message was sent ("Redirect to home phone failed: line busy"). The user did not respond at this point so the call is automatically forwarded to the default forwarding line (the cell phone). There is no answer from the cell phone so the call is re-forwarded to the default voice mail line (the work phone).

5 Discussion

The PAMMS architecture presents an agent-based approach to the enhancement of a universal personal numbering service. Though the UPN application is limited to telephony, voice-based communications and does not handle all mediums (e.g. email), it does provide the user with maximum flexibility when handling telephone calls and notification on non-telephony devices.

The PAAMS architecture itself has the adapters to manage other forms of information that include asynchronous multi-part messages like FAX, e-mail, and voice messages. The PAMMS model has allowed us to develop three different types of UPN services. The first used a PCA centered approach [8], the second deployed a UPN as an agent [9], and the third, presented here, uses the PCA to configure the UPN and the adapter agents to enhance the UPN's ability to communicate with non-standard telephony devices.

The current version of the system has several oustanding issues that will be addressed in the future. Currently, each agent is launched separately across four different computers. There is no ability to control the agents from one management platform. One proposed remedy is to employ mobile agents to enhance the deployment of customized services.

Another concern is the distributed ontology files for the service adapter agents. The agent's knowledge, in particular the knowledge used by an agent that behaves as a client to an adapter agent, should be maintained in the service adapter agent. The result of this is that one adapter agent need only load one JESS file. Client agents would be programmed by the exchange of knowledge when they first connect to an adapter agent.

6 Conclusion and Future Work

We have presented the Universal Personal Numbering System, an application that works within the framework of a Personal Agent Mobility Management System to provide remote notification, handling and redirection of telephone calls. The agent-based, modular architecture allows for maximum adaptability and extensibility on the part of the system operator. Users are able to manage their system both remotely and locally due to the multiple interfaces to the UPN application.

As the system moves away from a centralized PCA-based system to a decentralized system with customized adapter agents, we will be implementing multiple personalized adapter agents. This facilitates one system maintaining multiple PCAs where each will deploy their own set of adapter agents instead of sharing one multi-threaded adapter agent. It also cleanly separates the knowledge among the agents such that each agent only supports one user. The service adapter agents will be deployed by the PCA and have the ability to migrate to other host systems.

The current system has tackled the problem of isochronous communication (i.e. forms needing notification and connection). We are investigating the development of an XML-based universal messaging box to handle asynchronous forms of communications such as e-mail, voice mail and FAX.

The UPN and PAAMS Systems are just one facet of a network of new agent-based approaches in the field of unified messaging. At a core level, these systems provide a window to future developments in distributed-object and Java-centric technologies in personal communications management. At an interface level, we see further emphasis on minimum-hassle use on the part of the user and increasing ability to handle multi-modal messages from any sort of device.

References

1. Eckardt, T., Magedanz, T., Popescu-Zeletin, R.: A Personal Communication Support System based on X.500 and X.700 standards. Computer Communications. Vol. 20. (1997) 145–56
2. Schulzrine, H.: Personal Mobility for Multimedia Services in the Internet. In: European Workshop Interactive Distributed Multimedia Systems and Services IDMS '96. (1996) 143–61
3. Abu-Hakima S., Liscano R., and Impey R.: A common multi-agent testbed for diverse seamless personal information networking applications. IEEE Communications Magazine. Vol. 36(7). (1998) 68–74
4. Iida, I., Nishigaya, T., and Marukami, K.: DUET: Agent-based personal communications network. In: Int. Switching Symposium ISS'95. Vol. 1. (1995) A2.2
5. Friedman-Hill, E. J.: JESS, The Java Expert System Shell. Distributed Computing Systems, Sandia National Laboratories. Report SAND98-8206. (1997)
6. Labrou, Y. and Finin, T.: A Proposal for a New KQML Specification. University of Maryland, Computer Science and Electrical Engineering Report TR CS 97-03. (1997)
7. Liscano, R., Meech, J., and Impey, R.: Configuring a Personal Communications Agent. In: The Practical Application of Intelligent Agents and Multi-Agents PAAM'99. (1999) 423–436
8. Ber, P.: Pure Agent Messaging System. Seamless Personal Information Networking Group, Institute for Information Technology, National Research Council of Canada Internal Report. (1998)
9. Trzesicki, Z.: The Seamless Messaging System's TAPI Interface. Seamless Personal Information Networking Group, Institute for Information Technology, National Research Council of Canada Internal Report. (1998)
10. Gallagher, M.: UMTS: The Next Generation of Mobile Radio. IEE Review. Vol. 45(2). (1999). 59–63
11. Eckardt, T., Magedanz, T., Popescu-Zeletin, R., Schulz, M., Stapf, M.: Personal Communication Support in the TINA Service architecture - A new TINA-C Auxiliary Project. In: TINA'96 Conference. (1996). 55–64
12. Magedanz, T., Rothermel, K., Krause, S.: Intelligent Agents: An Emerging Technology for Next Generation Telecommunications? In: Infocom'96. (1996). 464–472
13. Messerschmitt, D. G.: The convergence of telecommunications and computing: What are the implications today?. Proc. of the IEEE. Vol 84(8). (1996) 1167–86

Abrose: A Co-operative Multi-agent Based Framework for Marketplace

Eleutherios Athanassiou[7], Peter Barrett[8], Delia Chirichescu[2], Marie-Pierre Gleizes[3], Pierre Glize[3], Dimitrios Katsoulas[4], Alain Léger[1], Jose Ignacio Moreno[6], Hans Schlenker [5]

[1]France Telecom CNET, BP59, 35512 Cesson Sévigné, France.
alain.leger@cnet.francetelecom.fr

[2]Infomures, Research Institutes for Informatics, Bucharest, Romania.
Delia@sysnetl.ici.ro

[3]IRIT-University Paul Sabatier, 31062 Toulouse, Cedex 4, France.
gleizes@irit.fr

[4]University of Athens, Athens, Greece.
Dkats@sftlab.ntua.gr

[5]Technical University of Berlin, DAI Laboratory , Germany.
Hans@cs.tu-berlin.de

[6]Universidad Carlos III de Madrid, Spain.
Jmoreno@it.uc3m.es

[7]DeutscheTelekom Berkom GmbH, Goslarer Ufer 35, D-10589, Berlin, Germany.
e.athanassiou@berkom.de

[8]Tradezone International Ltd., Zetland Bldg., Exchange Sq., Middlesbrough, UK.
Peter.Barrett@tradezone.co.uk

Abstract. The concept of Electronic Service Brokerage is based on the use of new information technologies so as to provide a service capable of facilitating and organizing the relationship between customers and offer providers. It covers the following main features: collaborative agents for self-organizing multi-agent systems, multi-agent system for knowledge representation of the market place domain, dynamic knowledge capture for domain knowledge evolution, ACL from Fipa.
In the first part we present the requirements for Abrose in using a Multi-Agent system for Brokerage. The second part presents briefly some agent tools in electronic commerce. The following three parts develop the architectural and functional choices in order to have the learning capabilities in Abrose.

1 Introduction

Electronic Commerce is based on the exchange of information between involved stakeholders, using the underlying telecommunication infrastructure. Business-to-Consumer applications enable providers to propagate their offers, and customers to find an offer which matches their demand.

A huge collection of frequently updated offers and customers, *the electronic marketplace*, reveals the task of matching the demand and supply processes in a commercial mediation environment.

In this paper, a framework for Electronic Brokerage Service based on the emerging agent technology is presented. The table below gives the expected functionality of Abrose, which is further developed in the text. The second part examines main aspects related to the current use of Multi-Agent systems for Brokerage. The third part presents the interaction principles between the Abrose components. In the fourth part, a typical example of Electronic Brokerage Service is detailed. The following part shows the self-organization mechanism underlying the learning functionality implemented in Abrose in order to take into account the high dynamicity of the Internet world.

1.1 Expected service elements

The Abrose system studied in this paper realises brokerage in the Electronic Commerce domain. In [Clurman,1997], the general commerce framework is synthesised in two parts: brokering and negotiation. They define brokering as the matching of buyers and sellers for the purpose of getting to a deal. The process of search is symmetric for a buyer or for a seller. The seller wants to know which buyers are potentially interested in his products. A buyer wants to find the best seller according to his criteria. A summary of applications, service elements for each actor and the basic technology main features within ABROSE are presented in the table below.

Applications	Electronic market places, offering Brokerage services
	• in the domains of Tourism , Telework and Business to Business Electronic Commerce
	• and some kind of One to One marketing, with intelligent support.
Service Elements for Customers	Simplified interactions for the customers
	• personalised assistance and notification for the customers
	• navigation and querying via assistance and unified view of the heterogeneous offers and demands
	• propagation of demands (" that's the work that travels to the people ! ")
for Content Providers	Simplified interactions for the customers
	• propagation of offers only to potentially interested customers
	• acquiring information of the customers real interests on the market in general and on the offers of the provider
for Market Place operation	Improvement of the quality of the mediation
	• flexibility to adapt to continuing changes of the demands and the offers
	• the experience acquired is made available to all customers (collaborative knowledge capitalisation and sharing, collective memory and recommendations,

	kind of Data Mining)
Basic Technology	**Intelligent (multi) agents technology allowing many features**
	• **autonomy** : e.g. for capturing dynamically the evolution of the environment
	• **co-operation** : e.g. for networking the expertise available in the broker domain
	• **delegation** : e.g. for user support in profiling the request and the return results
	• **own skill** : e.g. for achieving a task or a sub-task
	• **reasoning** : e.g. for evaluation of its own and others skills
	• **learning** : e.g. for evolving towards better performances by knowledge capitalization and sharing

2 Short overview on electronic brokerage

Due to the facility given by Internet and Intranet networks, the amount of available information becomes greater and greater. So, functionality is required to guide the users and the information or service providers through information transactions. One mechanism to provide this function is the information broker. The problem of the brokerage concerns how to retrieve and deliver specific information from the great amount of information available.

2.1 Recommender systems

The brokerage services enable users to browse through a directory or a classification until the user has found the appropriate information. Then he may provide attribute/value pairs until he finally gets the information he requires. A more flexible approach is now used by the way of personalised recommendations. Two great technical sets of recommender systems are designed :

- The first set uses an analysis based on natural language approach in order to select information to present to users. In return the users improve their profiles in evaluating information they received. We can find some system description in [ACM,1997], [Balabanovic,1997].
- The second set uses syntactic collaborative filtering. For example, Firefly allows users to get advice from other users for buying books and CDs. From the technical viewpoint, the service creates communities of common interest. Arachnid http://www-cse.ucsd.edu/users/fil/agents/ uses artificial life algorithms for information search. Agent based technologies are also used [Aoun,1996] or [Clurman,1997].

The operation of these user groups depends critically on a balancing between the number of users and providers, the view point of their interest and the algorithms used to decide similarity and dissimilarity between profiles. These grouping approaches reveal in fact the inability to manage a real dynamic users profile and to assume the true existence of a global functional adequacy as defined in [Piquemal,1996].

The dynamic is very high in the information brokerage because every time, new users or new information sources are autonomously created or removed in a distributed manner. Furthermore, the amount of information accessible through the Internet always becomes more and more numerous. These properties increases the difficulty to implement mechanisms to retrieve relevant information to a user.

2.2   Agents and Multi-Agent Systems in Brokerage

Multi-agent systems are computational systems composed of several agents capable of mutual and environmental interactions. The agents can communicate, co-operate, co-ordinate and negotiate with one another, to advance both their individual goals and the good (or otherwise) of the overall system in which their are situated [Demazeau,1998].

Software agents or softbots based on Artificial Intelligence, are developed to take into account the distributed nature of information space and the constant change of information content. Agents are software entities that operate semi-automously, performing operations on behalf of a user. They differ from classical tools in that they are personalised, continuously running and semi-autonomous [Guttman,1998].

Agents like Topic ™ (Verity ™), Internet Sleuth (MCCarnot), allow searching on the web in depth with a friendly interface. Autonomy (Cambridge University), is a search engine capable of dynamic reasoning suggesting similar notions to reach the greatest possible precision. For more information see http://www-sv.cict.fr/urfist
Some of systems like BargainFinder or Jango use content-based filtering methods to select merchants. The techniques are based on keyword search or are more complex in extracting semantic information from a document's contents.
The collective filters represent a new means to evaluate the resources on the Web.
- Metadata: Dublin Core Metadata and Warwick projects deal with automatic classification (http://www.dlib.org/dlib/july96/07contents.html) and define a set of metadata and methods to integrate them in the web pages: metatags (titles, authors, style, type of document)
- Boulder University has developed the program Harvest which indexes its own pages that are transmitted to the various search engines (http://harvest.transarc.com).
- cooperative filtering systems, like Group Lens, Phoaks, or Firefly http://www.firefly.com/ or Yahoo™ (at a less degree) are based on the favourite sites of groups of people sharing the same interests. They are based on feedback and ratings given by customers. Such systems carry the dangers of influencing the users (hiding marketing considerations under recommendations).
Many authors [Bothorel,1998], [Foner,1997], [Genesereth,1994] have developed agent applications in the field of Cooperative Information Systems. The basic assumption of these works is that an entity (for example a mediator) has the ability to link different terms or topics in order to find the right information source. In a dynamic world, this assumption must be replaced by some learning mechanisms about concept acquisition and relationships [Bollen,1996].

The main limitations found by Turner and Jennings [Turner,1997] are :
- They are less effective as the time passes because many of them rely upon index generators for their operations and the resources indexed are often outdated.
- They satisfy their user's requirements from a purely individualistic perspective. So if every user has its own softbot, then there would be millions of agents scouring the web. Furthermore, they don't co-ordinate their activity, they don't share information.

The conclusion is that sharing information and co-operating are necessary to increase utility, efficiency and scalability. The evolution would be in the sense that both users and information providers should be designed as agents. The multi-agent or agent technology are the most relevant to take into account this kind of application with no central control and high level of dynamicity.

Kasbah [Chavez,1996] is a Web based system where users create agents to negotiate for the purchase and sale of goods on their behalf. The selling agents are pro-active; they are not passive in their medium. The negotiation to reach an agreement between the seller and the buyer is based on constraints given to agents. The brokerage functionality is completely taken into account by the marketplace. It has to match up agents interested in the same goods. When a selling agent is created, the market place has to give to it a list of potential buyers agents and to inform all potential buyers of the existence of this new selling agent.

Bargain bot [Bassam,1996] is an electronic shopping agent based on multi-agent technology. Its specificity is a multi-thread, architecture, allowing for one search to create several sub-agents in charge of searching in a particular site. This parallelism diminishes the response time to the end user. It has prior knowledge of appropriate sites, knowing exactly where to locate its information. So the brokerage functionality is quite reduced.

Amalthaea [Moukas,1996] is a multi-agent evolving ecosystem where Information Filtering Agents and Information Discovery Agents co-operate and compete in a limited resource environment. The Information Filtering Agents are responsible for the personalization of the system and adapting to the interests of the user. The user gives a feed-back on the relevance of documents found to modify the behaviour of IFA. The Information Discovery Agents are responsible for finding information resources, handling them, finding and fetching the relevant information for the user. The technique used for adaptation is genetic algorithms.

The Abrose multi-agent based system [Gleizes,1999] is concerned with the brokerage functionality for the electronic commerce activity. The buyers and sellers are represented by pro-active, autonomous, cooperating agents. The originality of this broker is in using a collective memory to find a relevant agent and in the learning process which updates the knowledge of the collective memory at several levels. Furthermore, each agent has its own point of view of itself and of some others agents in the system.

3   The technology layer of Abrose

The Abrose platform V 1 is implemented in Java with JWS1.1.3, the communication between the Broker and the User domains is developed with OrbixWeb3.1(a Java implementation of CORBA from IONA). The brokers runs under Solaris 2.6. All the specifications have been made in UML under Rational™ Rose98™ . The client is a Personal Computer equipped of a standard browser. Netscape was chosen as the web browser of the system, because its Java VM fitted quite well our requirements.

3.1 Inter-modules communications

We will talk about communications between different modules in the broker domain (the 'server'), in the user domain (the 'client') and finally we will also consider inter-domain (client-server) communications and management communications. Several approaches have been taken depending on the domain.

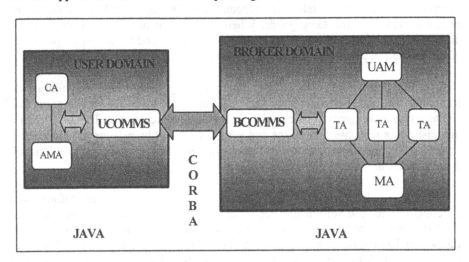

Figure1 : Communications between the main modules

The User Domain consists on a web browser (Netscape) running several Java modules (CA, AMA, ...) all of them downloaded from a http server. An invisible applet (no graphical interface) is first launched, and then the rest of the modules are run using independent windows (frames) if they require graphics (CA, AMA) or independent threads if they do not require so (UCOMMS).

All these modules use Java object reference to communicate between them, so they only need a method to exchange that reference in order to allow communications. Once these references are known, communications are as easy as executing the desired method of the referenced module.

UCOMMS module is responsible for communications with the broker domain, which will be explained shortly afterwards. Anyway, its communications with the rest of the modules of the user domain are through Java object references as well.

The broker domain communications are done in the same terms as in the user domain. Every module is an independent thread (BM, UAM, TA, BCOMMS, MA....) although there is no graphical interface.

BM is first launched, and it is responsible for starting the other modules and delivering references. As mentioned, modules can contact each other using the corresponding thread reference.

BCOMMS will be responsible for communicating with UCOMMS, and thread references are the way to contact the other modules of its domain.

Inter-domain communications between broker domain and user domain have been implemented using CORBA. A centralised approach has been chosen, so there are two modules (one for each domain) that will be responsible for sending and receiving information to and from the other domain. Whenever the user requires some information to be sent to the broker, the UCOMMS module will be called. BCOMMS would be the inter/domain communication module for the broker.

With this architecture, only UCOMMS and BCOMMS have to deal with CORBA, and communications can be more comfortably managed. The other modules just need a Java invocation (using their Java thread reference) to UCOMMS or BCOMMS in order to access the opposite domain.

3.2   Inter-agents communications

In the broker domain, TAs and MAs are autonomous agents (typically many hundreds which interacts by the way of a language act. A language act is neither an invocation method nor a procedure call. But this approach for interacting software is useful in an openness approach of systems where some unknown or unexpected events could occur. The main role of ACL is to give the ability to adjust autonomous activities. This is the main objective of some formalisms defined, such as ACL by FIPA [Fipa,1997]. The goal of the twenty communicative acts defined in ACL is to assume completeness, simplicity and conciseness without any presupposition on the social behavior of the agents.

Communicative acts for cooperative agents
The background work on the set of acts is to verify formally or experimentally that the current set of ACL is complete and coherent. But in the case of cooperative agents such as in Abrose (see part 5), all these acts can be simplified and exclusively performed by cooperative agents [Glize,1999] :
1. **Inform-ref.** This basic act is performed if the agent believes that the object corresponding to the definite descriptor is the one that is given, and does not believe that the recipient of the act already knows this.
2. **Query-ref.** The agent is unable to perform the previous inform-ref, but it believes that other agents are able to do it. The query-ref is exactly the initial

query-ref it receives sent to the set of believed relevant agents. This process is called relaxation in Abrose. A Query-ref occurs also when the agent is unable to perform inform-ref and also unaware of other relevant agents to perform it. In this case it creates a new query-ref directed to another agent on the upper level (such as a directory facilitator for brokerage).

3. **Failure.** A cooperative agent returns a failure only if all the previous acts cannot be performed.
4. **Not-understood**, in the other cases.

For a cooperative agent, returning directly a response of a given query or finding relevant agents able to give it, is included in the same basic and implicit behavior. Thus it seems that a cooperative agent performs high-level brokerage activities.

4. The Marketplace Metaphor

The ABROSE service paradigm is illustrated with the metaphor of a marketplace. When entering a physical marketplace, one realizes visually, in front of which shops people are queuing up, which restaurants are empty and which are always full, which magazines are bigger and more fashionable than others. When acting on an "abstract" marketplace, one may experience which goods create turnover on that special market, and which companies are successful.

In fact, marketplaces have a kind of "collective memory", the market "remembers" success and quality, and this "success" is visible to customers and providers. This collective memory is composed of individual experiences or recommendations received by other customers. This knowledge influences the decisions of customers and providers. The underlying idea of the ABROSE service is to introduce, model, represent and process this kind of "collective" and "individual" memory to electronic marketplaces. Users and Providers are enabled to use the knowledge of the marketplace, in order to improve the quality of the mediation for both parties.

4.1 Application in TradeZone

The following scenario is drawn from the business domain of Tradezone International Ltd (Business-to-Business Electronic Commerce https://www.tradezone.onyx.net). It corresponds to a supplier who wants to reach existing customers with individually tailored special offers. Priors fine arts supplies high quality paintings and prints to consumers and businesses. The company has a product catalogue on Tradezone. From time to time, the company is able to acquire art of specific interest, such as line drawings of old coal mining villages in northern England. Now that they have an on-line catalogue, they want to target specific previous customers with special information when this type of product is acquired. There are two specific actors and their corresponding Transaction agents inside the broker domain : the Glaxo company (called Glaxo) and Priors Fine Arts (called

Priors). The only steps explicitly done by customers are those corresponding inside the users domains (Glaxo and Priors) to the steps 1, 4, 7 and 9. All those communications and treatments are managed by the multi-agent systems of Abrose.

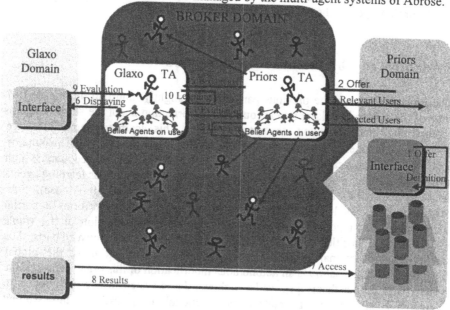

Figure 2 : Abrose activities during a brokerage scenario

1. Priors defines his offer using the available Interface. The aim of Abrose is to find customers who have previously shown an interest in this type of line drawing. For example, he might type: *"Mining, northern England, line drawings, charcoal, (artists name)"*

2. A message, which includes the offer, is sent to the Tradezone broker domain.

3. The Abrose broker creates a goal corresponding to this offer and analyses its beliefs in order to find the relevant customers in this particular application domain (being previous customers of Priors who have bought similar types of picture before). The Abrose broker has knowledge of all the experiences of the system. In particular it knows Glaxo because they have previously purchased through Tradezone. The Abrose broker presents to the Priors user the list of probably relevant customers with information about the previous buying.

4. Priors selects the desired customer(s) and sends the offer information to the selected customers. The User interface notifies the MAS about the experiences and selections made.

5. The offer is sent to all selected users by the Transaction Agent of Priors

6. The Glaxo TA which has received the Priors offer send it to Glaxo when it is logged in.

7. Interested by this subject, Glaxo connects to the Priors Provider using the address contained in the offer.

8. Priors accepts the transaction about paintings with Glaxo.

9. Satisfied by the result Glaxo sends to Abrose a positive evaluation.

10. This evaluation is learnt by the Glaxo Transaction Agent. This means that now Priors has relevant offers about paintings, and Glaxo customer is interested by.
11. The evaluation is sent to the Priors TA.
12. Priors TA updates the profiles users about itself and Glaxo.

5 Adaptive capabilities by self-organization

The process shown in the previous part could be slightly modified during the life-time of the system by taking into account new information about the skills of agents representing customers and content providers. This learning capabilities are necessary in an open and highly dynamic environment like the Internet. This kind of multi-agent systems capable of adapting their behaviour to take into account the high dynamicity of their environment is realised generally by autonomous learning agents or self-organisation between agents in the system (such as in Abrose). Self-organisation allows learning because each part of a system performs a partial function, and their composition gives the resulting global function of the whole system. The composition is the result of the internal organisation between parts; thus modifying this internal organisation is equivalent to transforming the global function. Self-organisation is an autonomous modification of this organisation by the system itself. In our theory [8], we have shown that a system where each part is in co-operative interactions, guarantees its functional adequacy, and this without knowing the global function of the system. To assume that the agents in a system are in co-operative interactions, it is necessary to give them the abilities to use uncooperative states to return to functionally adequacy. The notion of co-operation guides the learning activity which is performed in the three types of Abrose multi-agent systems.

The Abrose multi-agent architecture is a three-layer architecture, each layer is made of communicating agents (see part 3.2). An agent at a level is composed of co-operating agents at the sub-level endowed with a cooperative social attitude. This implies three main properties :

- ◆ Sincerity. If an agent knows that a proposition p is true, it cannot say anything different to others. This is a also main requirement in ACL.
- ◆ Willingness. All agents try to satisfy a received request if it is coherent with its own skills and the current state of the world
- ◆ Reciprocity. All the agents of the same society knows that itself and the others verify these main properties.

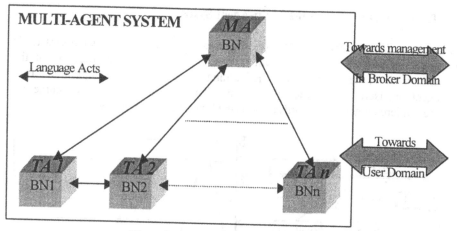

Figure3 : The multi-agent architecture

The top layer agents add in a totally incremental fashion new users and services located anywhere in the world. These agents called Mediation Agents (MA) are distributed on interrelated computers. MA can be added or suppressed inside the system, so avoiding the need to maintain any global view. The main difference of a MA and a shared memory is in the fact that a MA learns from the transactions done in the system and it does not require an update by the system designer. At the MA level, the right organisation is reached if all messages sent by any of them find a relevant receiver. When a MA believes that it is relevant for a received message, it passes it to TA composing it.

The second layer is made of Transaction Agents (TA), each customer or content provider in a domain is associated with a TA. TAs have beliefs on the skills of themselves and of other TAs which are hosted by the same MA. The TA is in charge of exchanging autonomously requests or answers with other TAs. The TA should also communicate with the customer or content provider it is associated with. At the TA level, the right organisation comes up when each TA is able to solve requests or to answer an offer received in messages and to know always partners able to help it. In order to find these partners, it uses its own beliefs.

All agents of the two previous types possess beliefs on themselves and on some agents belonging to the same level. These beliefs describe the organisation of the agent society [15]. They constitute the third layer of multi-agent system in Abrose called the Belief Network (BN) and are composed of Belief Agents (BA). The knowledge of an agent on others agents is continuously updated on the basis of the transaction flow. Thus, each layer improves its internal organisation in order to give efficiently and all the time relevant information. The next section of the paper explains this learning mechanism.

5.1 The learning process in the Multi-Agent System

A BA or a TA or a MA modifies the organisation of the system it belongs to, in changing its way of interacting with others. This implies a modification of the knowledge about the other agents. The learning concerns at this stage, the beliefs of an agent on itself and on the other agents. The BN has the form of numerous different interconnected keywords as described in the figure 4 below.

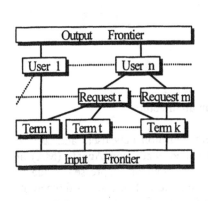

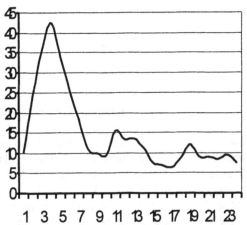

Figure 4 : The belief network organisation

Figure 5 : Example of network stabilization

A **term agent** represents a term in the belief network. It is filled by a term used during transactions between end-users and content providers. A **user agent** aggregates the beliefs about skills of a particular user or content provider (corresponding to his name) inside the belief network. A **request agent** is created when a user gives a feedback corresponding to an evaluation (good, bad, or dont care) of a previous transaction with another user. All the terms of this request are linked with a new request agent. The **frontier agent** gives the ability to all the others to perceive from and act to the external world of the belief network.

The self-structuring process

The principle of this structuring process is the following :
1. The propagation phase. The belief network is activated directly by the input frontier when a term is a member of the description, or indirectly by other agents which are entries for it. Each link is weighted, and when the sum of the activated entries are greater than a fixed threshold (1000 here), the agent is activated and then it sends a message to its acquaintances. At the end, all user agents activated are considered as relevant. The "understanding" of the belief network from the initial description corresponds to the aggregation of all the terms associated to active agents linked towards any activated user agents.
2. The analysis phase. Now each awaken agent analyses its internal state in order to see if some non cooperative situations occur. Two non cooperative states exist : Uncompetence : it is unable to use correctly the information received. The

result of these interactions is a modification of the weights between the agents implied in the communication (typically decreasing). Inutility : an activated agent does know any agent to send the result of its calculation.

3. The adaptation phase. After the analysis phase, concerned agents are informed about unexpected events which occur to the neighborhood. Now, they desire to solve these undesirable states in modifying the organization. This is realized firstly in modifying the weighted links (increment or decrement), and secondly in adding or suppressing links to agents. Thus the organization is concretely modified.

Asymptotic network structuring

The goal of the self-organization process in the belief network is to obtain a network similar to a thesauri without any presupposition about the domain at the conception phase. The belief network is totally empty before using it and the code managing it works without any rules about languages (English, French, German, Spanish,...) : this is the main characteristic of our learning process. The observation of the concrete behavior of a network leads to some results :

- A regular decrementation of non significant terms can be observed. This occurs because this term is very usual and used frequently without any particular relevance to any domain. When the weight of any agent falls to zero, the link between agents is suppressed.

- The positive evolution of weights associated to relevant terms in a given domain. All the terms – included articles, such as 'a", 'the" – have the same initial value, but the system is able to differentiate them. Moreover, if non significant terms disappear some specific terms stay stable, while generic terms increase.

- The number of agents involved in the network reorganization decreases during time, as shown in the example of the figure 5 (the abscissa indicates time, while ordinate corresponds to involved agents in restructuration). This is obtained because non significant terms are progressively suppressed from requests (after some times no uncooperative states are due to them), and also because the relevant terms of a domain are asymptotically linked with right weights (the number of their modification decreases).

The resulting consequence of these processes is the growing of generic terms inside the hierarchy where many more other specific terms will be linked. In this approach of learning, there is only one phase requiring user intervention, when he gives his feedback evaluation corresponding to the step 9 in the part 4.1. After, the network is able to adapt itself in order to reach an emerging organization between generic and specific terms such as a thesaurus. Moreover, this thesaurus is living because the new terms added in the network are obtained from the habits of end-users.

6 Conclusion

The Abrose platform V 1 is implemented, tested and validated. It is accessible remotely through the Internet (private access presently) for the Abrose members to demonstrate the system. Next step is now to specify and implement the final prototype for full demonstration of an Electronic Brokerage Service for real evaluations in three business domains (Tourism, Trading and Teleworking) as given in the part four. We currently implement new functionalities such as navigation for end-users inside a Belief Network in order to see the evolving representation of agents skills.

In addition, as active members of the FIPA consortium (Foundation for Intelligent Physical Agents), and as the FIPA97 and FIPA98 specifications are available [Fipa,1997], we partly evaluated and we reported results to FIPA [Glize,1999]. More precisely, on the need to implement the full set of ACL to the specific case of cooperative and self-organised multi-agent system, namely number of speech acts, protocols and ontology services. We have shown that only a small subset of ACL are needed to realize high level brokerage activities. In 1998, we have presented the specificity of the Abrose project to FIPA (9th meeting, Osaka, April, 98) and we acted as the editor of the "Methodology to build an ontology" (Annex B, FIPA98-part 10) published in fall 1998.

7 References

1. ACM – "Recommender Systems" - Special Issue of Communications of the Association of Computing Machinery – Vol40 N°3 - March 1997
2. Aoun Bassam, "Agent Technology in Electronic Commerce and Information retrieval on the Internet" http://www.ece.curtin.edu.au/~saounb/bargainbot , 1996
3. Balabanovic Marko - "An adaptive Web page recommendation service" - Proceedings of autonomous agents - 1997
4. Bollen Johan, Heylighen Francis - "Algorithms for the self-organisation of distributed, multi-user networks. Possible application to the future World Wide Web" Cybernetics and Systems'96 - R. Trappl (Ed.) - 1996
5. Bothorel Cécile - "Des communautés dynamiques d'agents pour des services de recommendation" - Proceedings of Journées francophones on Intelligence Artificielle Distribuée et Systèmes Multi-Agents - 1998
6. Camps Valérie, Glize Pierre, Gleizes Marie-Pierre, Léger Alain, Athanassiou Eleutherios, Lakoumentas Nikos - A Framework for Agent Based Information Brokerage Services in Electronic Commerce – EMSEC Conference – 1998
7. Camps V., Gleizes M.P. Glize P. 1998 Une théorie des phénomènes globaux fondée sur des interactions locales, 6ièmes journées francophones sur l'Intelligence Artificielle Distribuée & les Systèmes Multi-Agents, Éditions Hermès,
8. Chavez Anthony, Maes Patties, "Kasbah: An Agent Marketplace for Buying and Selling Goods", The First International Conference on the Practical Application of Intelligent Agents and Muli-Agent Technology, PAAM96, London, UK, April 96

9. Clurman Will, Foley Tim, Guttman Rob, Kupres Kristin, "Electronic commerce with software Agents, Electronic Commerce and Marketing on the Internet, Professors Tom Malone & John Little"

10. Demazeau Yves, Durfee Edmond, Georgeff Mike, Jennings Nick, Proceedings of the Third International Conference on Multi-agent systems, IEEE Computer Science, ISBN 0-8186-8500-X, Paris July 98

11. Etzioni Oren, Weld Daniel : "A softbot-based interface to the Internet", Comm. of the ACM, vol.37, 7, July 1994.

12. FIPA - "FIPA97 Specification Foundation for Intelligent Physical Agents" – Part2, Agent Communication Language - drogo.cselt.stet.it.fipa – October 1997

13. Foner L. N. – "Yenta : multi-agent, referral-based matchmaking system" – Proceedings on Autonomous Agents – 1997

14. Genesereth Michael R., Ketchpel Steven P : "Software agents" - Communications of the ACM, vol.37, 7 - July 1994.

15. Gleizes Marie-Pierre, Léger Alain, Athanassiou Eleutherios, Glize Pierre. "Abrose : Self-Organization and learning in Multi-Agent based Brokerage Services" ISN 99

16. Glize Pierre, Gleizes Marie-Pierre, Léger Alain - Brokerage communication in a cooperative multi-agent based mediation service : one example in Abrose - Foundation for Intelligent Physical Agent CFP6_016 - 1999

17. Guttman Robert H., Moukas Alexandros G., Maes Pattie. "Agent-mediated Electronic Commerce: A Survey. ", Knowledge Engineering review June 98.

18. Knoblock Craig A., Arens Yigal : "An architecture for information retrieval agents", AAAI Spring Symposium on Software Agents, Stanford, 1994.

19. Moukas Alexandros, "Amalthaea : Information Discovery and Filtering using a Multiagent Evolving Ecosystem", Conference on Practical Applications of Agents and Multiagent Technology, London, April 1996

20. Piquemal C, Camps V., Gleizes M.P, Glize P. : "Properties of individual cooperative attitude for collective learning"., Seventh European Workshop on Modelling Autonomous Agents in a Multi-Agent World (MAAMAW), 22-25 January 1996, Eindoven - The Netherlands.

21. Singh Munindar, Huhns Michael, "Challenges for Machine Learning in Cooperative Information Systems"- Lecture notes in AI - Vol 1221 - Springer-Verlag, 1997

22. Turner P.J, Jennings N.R, "On Scaleability of Information management Agents", Proceedings of the IT Conference (EITC-97), Brussels, Belgique, 1997

23. Wiederhold, Gio, "Mediators in the architecture of future information systems" IEEE Computer, Vol25 N.3, 1992

Author Index

Lecture Notes in Artificial Intelligence (LNAI)

Vol. 1545: A. Birk, J. Demiris (Eds.), Learning Robots. Proceedings, 1996. IX, 188 pages. 1998.

Vol. 1555: J.P. Müller, M.P. Singh, A.S. Rao (Eds.), Intelligent Agents V. Proceedings, 1998. XXIV, 455 pages. 1999.

Vol. 1562: C.L. Nehaniv (Ed.), Computation for Metaphors, Analogy, and Agents. X, 389 pages. 1999.

Vol. 1566: A.L. Ralescu, J.G. Shanahan (Eds.), Fuzzy Logic in Artificial Intelliegence. Proceedings, 1997. X, 245 pages. 1999.

Vol. 1570: F. Puppe (Ed.), XPS-99: Knowledge-Based Systems. VIII, 227 pages. 1999.

Vol. 1571: P. Noriega, C. Sierra (Eds.), Agent Mediated Electronic Commerce. Proceedings, 1998. IX, 207 pages. 1999.

Vol. 1572: P. Fischer, H.U. Simon (Eds.), Computational Learning Theory. Proceedings, 1999. X, 301 pages. 1999.

Vol. 1574: N. Zhong, L. Zhou (Eds.), Methodologies for Knowledge Discovery and Data Mining. Proceedings, 1999. XV, 533 pages. 1999.

Vol. 1582: A. Lecomte, F. Lamarche, G. Perrier (Eds.), Logical Aspects of Computational Linguistics. Proceedings, 1997. XI, 251 pages. 1999.

Vol. 1585: B. McKay, X. Yao, C.S. Newton, J.-H. Kim, T. Furuhashi (Eds.), Simulated Evolution and Learning. Proceedings, 1998. XIII, 472 pages. 1999.

Vol. 1599: T. Ishida (Ed.), Multiagent Platforms. Proceedings, 1998. VIII, 187 pages. 1999.

Vol. 1600: M. J. Wooldridge, M. Veloso (Eds.), Artificial Intelligence Today. VIII, 489 pages. 1999.

Vol. 1604: M. Asada, H. Kitano (Eds.), RoboCup-98: Robot Soccer World Cup II. XI, 509 pages. 1999.

Vol. 1609: Z. W. Ras, A. Skowron (Eds.), Foundations of Intelligent Systems. Proceedings, 1999. XII, 676 pages. 1999.

Vol. 1611: I. Imam, Y. Kodratoff, A. El-Dessouki, M. Ali (Eds.), Multiple Approaches to Intelligent Systems. Proceedings, 1999. XIX, 899 pages. 1999.

Vol. 1612: R. Bergmann, S. Breen, M. Göker, M. Manago, S. Wess, Developing Industrial Case-Based Reasoning Applications. XX, 188 pages. 1999.

Vol. 1617: N.V. Murray (Ed.), Automated Reasoning with Analytic Tableaux and Related Methods. Proceedings, 1999. X, 325 pages. 1999.

Vol. 1620: W. Horn, Y. Shahar, G. Lindberg, S. Andreassen, J. Wyatt (Eds.), Artificial Intelligence in Medicine. Proceedings, 1999. XIII, 454 pages. 1999.

Vol. 1621: D. Fensel, R. Studer (Eds.), Knowledge Acquisition Modeling and Management. Proceedings, 1999. XI, 404 pages. 1999.

Vol. 1623: T. Reinartz, Focusing Solutions for Data Mining. XV, 309 pages. 1999.

Vol. 1632: H. Ganzinger (Ed.), Automated Deduction – CADE-16. Proceedings, 1999. XIV, 429 pages. 1999.

Vol. 1634: S. Džeroski, P. Flach (Eds.), Inductive Logic Programming. Proceedings, 1999. VIII, 303 pages. 1999.

Vol. 1637: J.P. Walser, Integer Optimization by Local Search. XIX, 137 pages. 1999.

Vol. 1638: A. Hunter, S. Parsons (Eds.), Symbolic and Quantitative Approaches to Reasoning and Uncertainty. Proceedings, 1999. IX, 397 pages. 1999.

Vol. 1640: W. Tepfenhart, W. Cyre (Eds.), Conceptual Structures: Standards and Practices. Proceedings, 1999. XII, 515 pages. 1999.

Vol. 1647: F.J. Garijo, M. Boman (Eds.), Multi-Agent System Engineering. Proceedings, 1999. X, 233 pages. 1999.

Vol. 1650: K.-D. Althoff, R. Bergmann, L.K. Branting (Eds.), Case-Based Reasoning Research and Development. Proceedings, 1999. XII, 598 pages. 1999.

Vol. 1652: M. Klusch, O.M. Shehory, G. Weiss (Eds.), Cooperative Information Agents III. Proceedings, 1999. XI, 404 pages. 1999.

Vol. 1674: D. Floreano, J.-D. Nicoud, F. Mondada (Eds.), Advances in Artificial Life. Proceedings, 1999. XVI, 737 pages. 1999.

Vol. 1688: P. Bouquet, L. Serafini, P. Brézillon, M. Benerecetti, F. Castellani (Eds.), Modeling and Using Context. Proceedings, 1999. XII, 528 pages. 1999.

Vol. 1692: V. Matoušek, P. Mautner, J. Ocelíková, P. Sojka (Eds.), Text, Speech, and Dialogue. Proceedings, 1999. XI, 396 pages. 1999.

Vol. 1695: P. Barahona, J.J. Alferes (Eds.), Progress in Artificial Intelligence. Proceedings, 1999. XI, 385 pages. 1999.

Vol. 1699: S. Albayrak (Ed.), Intelligent Agents for Telecommunication Applications. Proceedings, 1999. IX, 191 pages. 1999.

Vol. 1701: W. Burgard, T. Christaller, A.B. Cremers (Eds.), KI-99: Advances in Artificial Intelligence. Proceedings, 1999. XI, 311 pages. 1999.

Vol. 1704: Jan M. Żytkow, J. Rauch (Eds.), Principles of Data Mining and Knowledge Discovery. Proceedings, 1999. XIV, 593 pages. 1999.

Vol. 1705: H. Ganzinger, D. McAllester, A. Voronkov (Eds.), Logic for Programming and Automated Reasoning. Proceedings, 1999. XII, 397 pages. 1999.

Vol. 1715: P. Perner, M. Petrou (Eds.), Machine Learning and Data Mining in Pattern Recognition. Proceedings, 1999. VIII, 217 pages. 1999.

Lecture Notes in Computer Science